D0872101

DATE DUE

~~DEC 0 00~~			
~~AUG 05~~			
~~MY C 5~~			

DEMCO 38-296

A DEFENSE OF GALILEO

APOLOGIÆ PRO
GALILEO

A Defense of Galileo
the Mathematician from Florence

Which Is an Inquiry as to
Whether the Philosophical View Advocated by Galileo
Is in Agreement with, or Is Opposed to, the Sacred Scriptures

by
THOMAS CAMPANELLA, O.P.
of Calabria

Translated with an Introduction and Notes
by
Richard J. Blackwell

UNIVERSITY OF NOTRE DAME PRESS
Notre Dame   London

R

Library of Congress Cataloging-in-Publication Data

Campanella, Thomas, 1568–1639.
     [Apologia pro Galileo.   English]
     A defense of Galileo, the mathematician from
Florence, which is an inquiry as to whether the philo-
sophical view advocated by Galileo is in agreement
with, or is opposed to, the Sacred Scriptures / Thomas
Campanella ; translated with an introduction and
notes by Richard J. Blackwell.
          p.   cm.
     Translation of: Apologia pro Galileo.
     Includes bibliographical references.
     ISBN 0–268–00869–8 (alk. paper)
     1. Galilei, Galileo, 1564–1642. 2. Copernicus,
Nicolaus, 1473–1543. 3. Religion and science—
Early works to 1800.
     I. Blackwell, Richard J., 1929–   . II. Title.
QB36.G2C3213   1994
215'.2—dc20   93–8497
                                              CIP

♾The paper used in this publication meets the minimum requirements of
the American National Standard for Information Sciences—Permanence of Paper
for Printed Library Materials, ANSI Z39.48-1984.

FOR ANNE, RICHIE, AND DANIEL

"And may you live to see your children's children
unto the third and fourth generation."

Psalm 128:5
as in the Nuptial Blessing

In my judgment . . . it is not possible to prohibit Galileo's investigations and to suppress his writings without causing either damaging mockery of the Scriptures, or a strong suspicion that we reject the Scriptures along with the heretics, or the impression that we detest great minds.

Defense, last page

If a scandal arises from the truth, it is better to let the scandal occur than to abandon the truth.

Gregory the Great,
as quoted by Campanella

CONTENTS

PREFACE

espite extensive study of the Galileo affair in recent years, there are still some important documents in the case which have received little or no attention, at least in the English-speaking world. One such treatise is Tommaso Campanella's *Apologia pro Galileo, mathematico florentino*, first published in 1622 but written early in 1616 at the same time when the Catholic Church was in the process of deciding on its fateful judgment to condemn Copernicanism.

There are two reasons for this neglect. First, for inexplicable reasons, Antonio Favaro did not include Campanella's *Apologia* in his *Le Opere di Galileo Galilei* (20 vols., Florence: G. Barbèra, 1890–1909), which has since served as the basis for subsequent studies of the documents relevant to the Galileo affair. Second, and perhaps more importantly, the only English translation to date of Campanella's treatise, by Grant McColley (*The Defense of Galileo* in *Smith College Studies in History*, vol. 22, nos. 3–4, 1937), is universally recognized as so inaccurate as to be unuseful not only for scholars but even for the general reader.

The purpose of the present volume is to establish an awareness of Campanella's role in the Galileo affair by finally making available a reliable and well-documented English version of his *A Defense of Galileo*, in which the theological dimensions of the dispute receive perhaps their clearest presentation.

On the other hand, two excellent Italian translations of the *Apologia* have been published in recent years. The first is Luigi Firpo's *Apologia di Galileo* (Turin: UTET, 1968); the second is Salvatore Femiano's *Apologia per Galileo* (Milan: Marzorati Editore, 1971). The former also contains a facsimile reprint of the original 1622 Latin edition; while the latter contains a modern Latin printing with corrections of the typographical errors in the original edition. We are greatly indebted to both of these scholars and to Grant McColley for their detailed notes which were invaluable in tracking down and identifying the names, titles, and references mentioned by Campanella in the Latin text.

Our English version of the *Defense* is based on the 1622 first edition as corrected for typographical errors. No autograph copy of Campanella's original Latin text is known to exist, but fortunately the text of Tobias Adami's 1622 Frankfurt edition is in good condition. As a result no major textual problems arise for the translator; a few minor variations are mentioned in our notes.

In preparing the translation we have focused on both accuracy and readability but hopefully have never compromised the former for the latter. The task was complicated by the fact that Campanella's Latin in this treatise is extremely compressed and almost devoid of stylistic flourishes. Many new paragraph divisions have been added in the translation to break up the author's excessively long paragraphs so that the text will be more readable.

Everything in the translation is contained in the original Latin text except for the occasional brief material enclosed within square brackets. Almost all of these additions consist of more specific biblical references than Campanella himself gives (he usually quotes Scripture by chapter but not by verse, and his memory is sometimes inaccurate about chapter numbers.) This material has been added by the translator in square brackets at the appropriate places primarily for ease of reference and to avoid an excessive number of endnotes.

In the endnotes we have attempted to identify as many of Campanella's references as we could, although we clearly have not succeeded in finding all of his specific sources. The fact that he wrote largely from his own prodigious and amazingly accurate memory suggests that there may be some errors in the references which remain unidentified. Also in the endnotes we have identified proper names where first mentioned by Campanella by providing full names, alternate names, dates, and at least a brief set of remarks to enable the reader to find more information in other biographical sources.

Finally we wish to express our thanks to James Langford, Director of the University of Notre Dame Press, both for his original suggestion that we undertake this translation and for his editorial help and encouragement along the way. We also must express our appreciation for the technical and textual assistance received from the professional librarians at the Pius XII Memorial Library of Saint Louis University and especially to Dr. Charles Ermatinger, the Director of its Vatican Microfilm Library.

INTRODUCTION

In 1616 Galileo failed in his efforts to dissuade the Catholic Church from condemning the newly revived sun-centered view of the world as "false and completely contrary to the divine Scriptures." At that time only two persons in the Catholic world were courageous enough, or foolish enough, to speak out publicly in defense of Galileo. Only two. Both were priests. Both paid a heavy price.

The first support for Galileo came from a relatively unknown Carmelite theologian, Paolo Antonio Foscarini, who wrote a booklet which attempted to produce the needed reconciliation between Copernicanism and the Bible. Within one year this book was "completely prohibited and condemned" and was placed on the *Index*. Foscarini died before he could complete the writing of a rejoinder which he had planned. His publisher in Naples abandoned his business and family in the hope of avoiding prosecution for violation of the publishing laws, but eventually he was arrested by the Inquisition, found guilty at a trial, fined 100 ducats, and later finally was pardoned under a promise of silence.

Galileo's second defender was the famous renegade Dominican, Tommaso Campanella, a man of genuine genius and imagination, who at that time had already been in the prisons of the Inquisition for seventeen years under charges of heresy and conspiracy. His carefully reasoned book, which surprisingly enough was written at the invitation of a cardinal who was a member of the Congregation of the Holy Office then sitting in judgment on the case, was effectively muffled in Rome and had little or no influence on the actual course of events. But at least the best arguments of the defense had been heard in Rome. This book was finally published six years later, in Germany rather than in Italy, as its author languished in the same prison at the midpoint of another twelve years under the original charges.

Tommaso Campanella's extraordinary defense of intellectual freedom in the Galileo affair is the subject of this book. At the time he had only occasionally met Galileo many years earlier, and their acquaintance since then had been limited to a few exchanges of letters. For obvious reasons Galileo in turn chose not to develop a close friendship with a prisoner of the Inquisition. Although Campanella clearly had a great admiration for Galileo, personal loyalties were not his main motive in this matter. Even more significantly Campanella himself believed that the heliocentric model of the universe was doubtful and not proven. Nevertheless he maintained an unshakable conviction that, for the benefit of the Church itself, Galileo should be permitted to develop, debate, and publish his ideas freely. It was the welfare of the Church, and not that of science nor of Galileo himself, which was Campanella's main concern. Written in prison under the harshest of conditions and at great personal peril, his book, *A Defense of Galileo*, was a plea for intellectual freedom which far transcends the specifics of Galileo's confrontation with his Church.

One can hardly imagine a more startling or more dramatic set of circumstances than those surrounding Campanella's book. Who was this unusual man, and how did he become involved in such an extraordinary way in the Galileo affair?

THE TRAGIC LIFE OF A RECALCITRANT SPIRIT

Giovanni Domenico Campanella was born on 5 September 1568 in the town of Stilo in the region of Calabria in southern Italy. His mother's name was Caterina Martello; his father, Geronimo, was a shoemaker. His parents were illiterate and poor. Although he came from quite humble origins, in his writings, including the *Defense*,[1] Campanella often spoke with great pride of his native Calabria, which had long ago been part of Magna Graecia, the home of Empedocles and the Pythagoreans. In his later life he saw himself as reviving and continuing the empirical and mathematical study of nature which had been initiated by his famous compatriots more than twenty centuries earlier.

In 1582, at the young age of thirteen, he joined the Order of Preachers. He remained a Dominican for the rest of his life. He was attracted to the order in part by his reading about the lives of the famous Dominicans Albert the Great and Thomas Aquinas. Subsequently he adopted the religious name "Tommaso," by which he has been known ever since. He may well also have seen that, because of his humble origins, entering religious life was the only avenue open to him to acquire an education rather than follow the life of a craftsman like his father.

Despite the many hardships and sufferings which awaited him in that religious world, his loyalty to the Dominican Order and to the Church never wavered throughout his life. His persistent advocacy for educational and religious reform of both of these institutions increased rather than decreased his commitment to the life of a Dominican within the Church. In fact the strength of his loyalty may have made him seem even more dangerous in the eyes of some of his enemies in later life.

Although his childhood education was quite inferior,[2] Campanella's extraordinary intellectual talents and the wide range of his interests—from theology to philosophy, politics, and poetry—became apparent in his early years. For example, by the time he joined the Dominicans he not only knew Latin but was widely read in a variety of Latin authors.

Of equal significance is the fact that he possessed

what we would now call a "photographic memory," which enabled him to retain, often *verbatim*, much of what he read. As he himself reports in one of his letters, "When I read a book, I am so affected by its reading that the very words and the content of it remain impressed into my memory almost indefinitely."[3] As we shall see, his prodigious memory served him very well during his many years of imprisonment by enabling him to continue his writings even though he was forced to work with few, if any, books actually in hand.

Campanella's basic intellectual formation took place in the first ten years or so after he joined the Dominicans. During that period, 1582–1592, he lived at several monasteries in Calabria and in Naples, as he pursued the prescribed sequence of studies in philosophy and in theology. The former included the usual heavy dose of Aristotelianism, which he came to strongly reject and openly dispute with his teachers. These disputes grew in intensity over the years, and his persistence created enemies among his brother Dominicans. To make matters worse Campanella began to read widely in the writings of other ancient philosophers and scientists, and he began to develop an interest in magic and the occult. The latter was a particularly sensitive topic as far as Catholic orthodoxy was concerned.

In regard to his anti-Aristotelianism, Campanella's point of view was that Christian schools should prefer the teachings of the Scriptures and the Fathers of the Church over the pagan Aristotle, who denied the doctrines of creation, providence, and the immortality of the soul. This anti-Aristotelian sentiment remained a prominent feature of his writings for the remainder of his life. It is found in many places in the *Defense*, since Aristotle was also the father of the geocentric model of the universe challenged by Galileo.

Furthermore, especially in his early years, Campanella strongly favored direct observation of nature rather than the prevailing tendency to turn first to the writings of

Aristotle, as the best way to understand the natural world. Such an empiricist thrust was just beginning to receive expression in those years as a harbinger of the modern scientific approach. Because of this empiricist disposition, when in 1588 Campanella discovered Bernardino Telesio's (1509–1588) recently published *De rerum natura iuxta propria principia* (1586), he immediately became an advocate of Telesian philosophy. His attempt to meet with his newly found intellectual hero failed, however, since Telesio died later that same year, and Campanella arrived only in time for the funeral. His support for Telesio's philosophy was the subject of Campanella's first major treatise, *Philosophia sensibus demonstrata* (written in 1589, published in 1591), whose very title expresses the empirical gounding needed for philosophy.

Telesio rejected Aristotelian hylomorphism and its notion of primary matter as pure potency. In its place he developed a natural philosophy based on two principles: (1) concrete, actual, sensible matter, which is passive and the same everywhere in the universe, and (2) two opposed active forces, heat and cold, which are the causes of all physical change. The interaction of these factors constitutes the structure and dynamics of the physical world, including plants, animals, and the human body and spirit.[4]

As far as cosmology is concerned, Telesio maintained that the earth is at the center of the universe and is cold, motionless, and dark; while the hot and luminous sun revolves swiftly around the earth. It is highly likely that Campanella adopted this Telesian version of geocentricism in his early years and still maintained it as probably true when he wrote the *Defense*.[5] However by then (1616) he also granted some reasonable degree of probability to Galileo's heliocentric theory, although he clearly did not regard the latter as having been proven.

Later in life Campanella went beyond Telesian natural philosophy by developing a full-blown theological metaphysics. In this account all things, both God and creatures, both the immaterial and material worlds, are comprised of three principles or "primalities": power, knowledge, and love. In God this is the Trinity. In creatures, which are diluted by non-being as well, the three "primalities" are found in all things but in a wide variety of degrees and combinations, resulting in a version of panpsychism. This view is behind Campanella's remarks in the *Defense* that human reason participates in the divine reason of Christ.[6]

One result of this was that Campanella's original emphasis on empirical method became somewhat muted by a type of Scholastic metaphysics which took precedence over his natural philosophy. From this perspective Campanella seems to look less toward the future, where modern science waited to be born, and to look rather toward the past of late medieval Scholasticism, where metaphysics reigned. On the other hand he might well have become much more of an experimentalist in his later life, had his long years in prison not prevented him from carrying out such work and thus forced him to think more theoretically. At any rate, Campanella's early intellectual formation gradually grew into a complex worldview which combined Telesian philosophy of nature with his own metaphysics of power, knowledge, and love.

The decade of the 1590s saw two major changes in Campanella's life. First, at the end of 1589 he created a scandal when he left the monastery and went to live in private quarters in Naples. Two and a half years later he left Naples and lived in several Italian cities to the north until 1598. Second, his confrontations with the authorities of his order, the Church, and the Spanish state began. Numerous charges of heresy against the Church and conspiracy against the state, along with several temporary confinements, culminated in his long-term imprisonment beginning in 1599. This is a very complex story, but we have space to provide only the main developments.

In May of 1592 Campanella was denounced for heterodoxy by an anonymous Dominican brother and had to stand trial before the order. The specific charges are unclear but apparently included his anti-Aristotelianism and his association in Naples with Giambattista Della Porta and other advocates of magic and the occult.[7] He was ordered to return to a monastery in Calabria, to adopt Thomistic rather than Telesian philosophy, and to perform various penances.

Instead he went to Rome and later to Florence in the hope of gaining an academic appointment from Grand Duke Ferdinand I. Still later he traveled on to Bologna (where his papers were stolen and sent to the Holy Office in Rome), then to Padua, and finally to Venice (where he hoped to publish some of his books). In 1594 his writings were seized again, and at the Holy Office in Padua he was charged, among other things, with criticisms of the organization and dogmas of the Church. The case was later transferred to Rome, where he was imprisoned, tortured twice, and finally found guilty late in 1596 of "strong suspicion of heresy." His sentence was to make a public abjuration at the Dominican Church of Santa Maria sopra Minerva and to be confined at the monastery of St. Sabina.

On 5 March 1597 he was charged again with heresy, this time by a man facing execution in Naples. He spent the remainder of that year in prison, being released only on condition that he be confined to a monastery. He thus found himself back at the Dominican house in his native Stilo in 1598. During these tumultuous years Campanella somehow found time and energy to continue his writing, but many of these manuscripts have been lost or destroyed.

During these sojourns Campanella's one peaceful year, 1592–93, was spent at Padua, where he developed an acquaintance with Galileo, whom he always later considered to be his friend. Unfortunately no documentation is known to exist which would tell us how the two men related to each other during that year. It has been suggested[8] that they first met late in the fall of 1592 when Galileo, newly arrived in Padua from Florence, personally delivered a letter for Campanella at the Convent of St. Augustine, where Campanella was staying. In the letter the Grand Duke politely rejected Campanella's request for an academic appointment.

Harsh as his personal life was in the 1590s, Campanella's really serious troubles with the authorities were just beginning. By this time his social, political, and religious views had developed to the point that he perceived a strong need to overcome corruption and injustice by means of a complete reformation of both church and state, which would in turn usher in a new golden age. He advocated a universal and communistic church-state, in which all temporal and spiritual power would be vested in one monarch, the pope, who would preside over an ideal, rational society under the one true faith. He spoke out publicly and forcefully for this plan, which became his lifelong dream.

One immediate consequence of this would have been the overthrow of Spanish authority in Naples and Calabria. A group of Calabrian conspirators developed, although it is hard to say how active Campanella was in the group beyond clearly being the source of its theoretical justification. At any rate, the plot failed, and Campanella was arrested on 6 September 1599, along with many of his associates.

He was imprisoned in Naples for what would turn out to be the next twenty-seven years.[9] He was charged with both heresy and conspiracy. This necessitated two trials, since the former charge was the concern of the Holy Office, while the latter charge was an offense against the Spanish authorities. His trials and interrogations dragged on for three years, during which period he was subjected to torture four times, including the horrendous "veglia" on the last occasion. It should be noted that at that time the purpose of torture was not to inflict punishment for admitted crimes but to induce the prisoner to confess. Somehow Campanella managed not to confess even under these dreadful circumstances.[10] At one point he feigned madness, which may have saved his life. Finally on 13 November 1602 he was sentenced to life imprisonment "with no hope for liberation."

During these long years in prison Campanella never ceased his appeals for help from various prominent potential benefactors. As his circumstances changed, he was moved between various prisons in Naples, some worse than others. His severest punishment was two confinements to the dungeons, in 1604–1608 and in 1614–1618. Ironically it was during this latter period that he wrote his *A Defense of Galileo* (1616).

At times he was allowed to receive occasional visitors and to maintain some correspondence with persons in the outside world. Most astonishingly he wrote voluminously during these years, working under the worst possible conditions, with few or no books and depending on his prodigious memory for sources. He must have been allowed at least some limited access to books, contrary to what some studies of Campanella say, since he refers to many books published after his incarceration, as can be seen in the *Defense*. For example, nine months after the publication of Galileo's *Sidereus nuncius* in March of 1610, Campanella received a copy indirectly from Galileo, who asked for his evaluation. Campanella "devoured" the book in two hours, and on 13 January 1611 he sent Galileo an enthusiastic letter[11] in response, but not in agreement, since the new astronomy conflicted with his own Telesian geocentric cosmology.

After what must have seemed an eternity, Campanella's special pleadings to the religious authorities were finally effective, and the Spanish government released him from prison on 23 May 1626. One month later he was arrested again by the Holy Office and sent to the prison of the Inquisition in Rome on charges of heresy in his writings. He was not able to overcome these complaints until 11 January 1629, when he finally regained his freedom. Thus his ordeal ended.

The next five years were spent relatively quietly in revising his writings and preparing them for publication. During that period Campanella lived in Rome at the Dominican Convent of Santa Maria sopra Minerva, which ironically was also the scene of the hearings in Galileo's trial in the spring of 1633. What passed through Campanella's mind as he lived at that time in the same building in which his friend Galileo was put to the test?

The previous summer Campanella had boldly volunteered himself to undertake Galileo's defense by requesting that he be appointed a consultor to the special commission which was just beginning its investigation of the orthodoxy of Galileo's *Dialogue Concerning the Two Chief World Systems.* This daring suggestion was menacingly rejected by Niccolò Riccardi, O.P., Master of the Sacred Palace, who remarked that this was impossible since Campanella was himself guilty, "since he had written a similar work which had been prohibited." This refers to the *Defense*,[12] whose sale was prohibited in Rome immediately after it was published.

Finally in the fall of 1634 word arrived in Rome that one of Campanella's former associates had been arrested in Naples on the familiar charge of conspiracy. Fearing personal implication, and for once following the advice of the pope, he left Rome in disguise on October 21 and went to Paris, where he was welcomed by Cardinal Richelieu. Continuing his writing and editing work, he lived the last years of his life quietly at the Dominican house on the rue St. Honoré in Paris, where he died on 21 May 1639.

COPERNICANISM AND THE CATHOLIC CHURCH

On the first page of the *Defense*, as well as in its subtitle, Campanella stated the central theme of his treatise as directly as he could. The question to be investigated was "whether the philosophical view advocated by Galileo is in agreement with, or is opposed to, the Sacred Scriptures."[13] To understand this issue more fully, it is necessary to ask a few further questions about the two parts of Campanella's stated theme. First, what was Galileo's "philosophical view," what had been its previous history, why did he advocate that view, and was he able to prove that it is true? Second, in that day what constituted agreement or disagreement with the Scriptures, how was that to be decided, and why was this agreement or disagreement considered to be so important?

Galileo's "philosophical view" was, of course, the heliocentric model of the universe, i.e., the view that the stars, the planets, and the earth with its moon all revolve around the sun, which in turn is located in the center of the universe. This is a very old theory which had been developed and defended by some of the ancient Greek astronomers, as Campanella points out. But it had fallen by the wayside, even in ancient times, when Aristotle's geocentric views about the heavens came to dominate Western thinking.

In briefest outline the Aristotelian model of the universe was as follows. At the center is located the earth which is motionless, i.e., it does not move through space, nor does it revolve on its own axis. The space from the center out to the moon is composed of the four elements: earth, water, air, and fire, which transform into each other, and which compose the changeable bodies of common experience in the terrestrial world. From the moon on outward to the end of our finite universe, there exists a fifth type of immutable matter, called the quintessence or "fifth essence," which undergoes only uniform and circular local motion in space. This matter is structured into a series of perfectly concentric spheres, some of which separately contain the moon, the sun, the five visible planets, and the fixed stars. As these solid spheres rotate at various angles and at individually different but constant velocities around the earth at their center, all the visible phenomena of the celestial world are thereby produced. These bodies and their motions are eternal according to Aristotle. The Greeks, of course, had no way to measure astronomical distances, so they assigned no specific dimensions to this finite universe as a whole. However they clearly considered it to be very small by modern standards, probably no larger than what we now know is the diameter of the earth's orit around the sun.

For about twenty centuries this view of the universe was almost universally accepted in the West. And there were good reasons for this. It fit very well with the commonsense view of the heavens, i.e., that they rotate daily from east to west but with minor variations, just as the theory says. This theory was able both to predict and to explain such celestial phenomena rather effectively. It was also a very adaptable model, since slight variations of the directions and velocities of the motions of the spheres, or even the addition of new spheres, could be and were introduced to account for new or for more accurate observations. Any theory which could last for so many centuries had to have a good deal of both empirical and theoretical support behind it.

But eventually it ran into some serious complications. One of these was the reliability of the calendar. In the Julian calendar, introduced by and named after Caesar, the length of the solar or tropical year[14] is 365.25 days (365 days plus February 29 every fourth year). Actually the tropical year is slightly shorter (365.24220 days, or eleven minutes and fourteen seconds less). By the beginning of the sixteenth century this discrepancy amounted to ten days, and it had become clear that the calendar needed correction.[15]

This was also of concern to the Church because of its need to determine religious holydays; for example, Easter is the first Sunday after the first full moon after the vernal equinox, so the exact date of the equinox was required. At the Fifth Lateran Council (1512-1517) this matter was discussed, with the result that Nicolaus Copernicus was asked to undertake the needed calendar reform. He appreciated the full complexity of the problem and decided for this and other reasons that the accepted Aristotelian-Ptolemaic astronomy of the day would need a complete reexamination.

The net result was the publication in 1543 of his monumental *De revolutionibus orbium coelestium*. Most of this long treatise is a highly technical mathematical presentation of basic astronomy, but in the first eleven chapters of Book I the long-forgotten heliocentric model of the universe was revived and presented as the needed reform in astronomy. In the simplest terms Copernicus merely exchanged the places of the sun and the earth (along with its moon) in Aristotle's model, with the result that the sun now lies motionless at the center around which the spheres revolve as before, except that the earth now rotates daily on its axis and annually around the sun in the third sphere, which is located between Venus and Mars. The scientific revolution began with this seemingly nondisruptive change, for the latter entailed considerable further consequences beyond what Copernicus had done; for example, a new terrestrial physics was now needed for an earth in motion.

It is significant to note that from the beginning this Copernican revolution was seen as potentially in conflict with the Bible, which usually speaks in the language of the commonsense geocentric model. This is evident from the fact that Andreas Osiander, who saw the book through the press as Copernicus lay dying, added an anonymous preface which advised the reader that the new heliocentric model was introduced only to simplify the mathematical calculations and was not to be taken as physically real. This "instrumentalist" resolution of the potential science-religion conflict was widely used in the generations to follow.

For the remainder of the sixteenth century the new Copernican astronomy won only a small, but slowly increasing, number of advocates, even among the professional astronomers of the day. But the picture changed considerably after the turn of the century. In the period between 1610 and 1613 Galileo succeeded in improving on the telescope (which had been invented in Holland in 1609) and in making a series of startling discoveries as he turned his "optical tube" to the heavens. He was able to see, among other things, four of the moons of Jupiter, the irregularity of the surface of the moon, the fact that Venus displays a set of phases, as does the moon, the existence of spots on the sun, and a vastly larger number of fixed stars.

These phenomena clashed in various ways with some of the specifics in the Aristotelian view of the world, which thereby became threatened. Galileo became personally convinced of the truth of heliocentrism, but he knew that he did not have a conclusive proof of that theory. His best piece of evidence, the phases of Venus, proved, contrary to Aristotle, that Venus revolves around the sun. But this did not determine whether the sun or the earth is at the center. The issue was further complicated by a third astronomical model introduced by Tycho Brahe; namely, that the planets and stars revolve around the sun, which in turn revolves around an earth at rest at the center. Tycho's geocentric model also fit the new observations rather well.

From the middle of the seventeenth century and beyond the Copernican heliocentric astronomy became almost universally accepted, and its conclusive proof came finally in the 1830s when the parallax of the fixed stars, predicted by Galileo, was finally observed with improved telescopes. But it is important to note that, despite his personal conviction of the truth of heliocentrism, the conclusive proof which Galileo sought throughout his lifetime always eluded him. This absence of a definitive proof weighed heavily on the minds of the Churchmen, especially Cardinal Robert Bellarmine, who played a central role in the Church's reaction to the new astronomy.

This brings us to the other part of Campanella's central theme in the *Defense*; namely, what were the views at that time about agreement or disagreement with the Scriptures, especially in relation to the new discoveries and theories appearing in astronomy? In Galileo's day the attitude of the Church on this question was determined primarily by the decisions made at the Council of Trent (1545–1563).[16] The central development was the decree adopted at the Fourth Session (8 April 1546) regarding the interpretation of Scripture. The relevant passage of that decree reads as follows:

> Furthermore, to control petulant spirits, the Council decrees that, in matters of faith and morals pertaining to the edification of Christian doctrine, no one, relying on his own judgment, and distorting the Sacred Scriptures according to his own conceptions, shall dare to interpret them contrary to that sense which Holy Mother Church, to whom it belongs to judge of their true sense and meaning, has held and does hold, or even contrary to the unanimous agreement of the Fathers, even though such interpretations should never at any time be published. Those who do otherwise shall be identified by the ordinaries and punished in accordance with the penalties prescribed by the law.[17]

To understand how Galileo and heliocentrism came to be condemned by the Catholic Church, it is necessary to have a good grasp of this decree. It asserts, first, that the Church (which in this case means the pope and the bishops) reserves to itself alone the power and the responsibility to determine what is the correct meaning of the Scriptures. No one else, including the lower clergy, has this role. Second, no Catholic may interpret the Bible contrary to what the Church has determined it to mean, or contrary to the unanimously agreed interpretation of the Fathers of the Church. Because of this, Campanella spends a great deal of time in the *Defense* quoting Scripture and the Church Fathers and theologians. He does this in an attempt to show what is or is not the accepted meaning of Scripture according to the Church and the early Fathers. Although such proof from quoted authorities may seem unusual or overdone to the modern reader, it was the accepted mode of theological argumentation of that day for the reason just mentioned.

Third, those who interpret the Bible differently than indicated are to be identified and legally punished for heterodoxy under Church law. Both Galileo and Campanella suffered these consequences. Fourth, the decree quoted above is expressly limited to "matters of faith and morals." This is a technical expression referring to dogmas defined by the Church as revealed truths and to various customs and practices in the Church, for example, the administration of the sacraments. Other information in the Scripture which does not fall under this category of "faith and morals," for example, that Tobias happened to own a dog, is apparently not governed by this decree.

It should be noted that the decree from the Fourth Session of Trent does not say anything about what principles of exegesis are to be used in the interpretation of Scripture. From the earliest centuries the Catholic Church was pluralistic on this point and had developed over time complex distinctions concerning the spiritual, literal, and metaphorical levels of meanings in the Bible. A strict biblical literalism has never been the dominant point of view in the Catholic Church. However, as the decades passed after Trent, there was an increasing emphasis on the literal meaning of the Scriptures, primarily as a tool to be used against the Protestants in the theological debates of the Counter-Reformation.

A good example of this is Cardinal Bellarmine, the most prominent theologian of his day and theological adviser to several popes. Throughout his life he believed personally that the cosmos exists as it is literally described in Genesis. In his later years, when he evaluated Galileo's heliocentrism in relation to the Scriptures, he even extended the Decree of the Fourth Sesssion of Trent beyond "faith and morals" to everything that is said in the Bible.[18] His argument was simple. The Scriptures are the written word of God. Hence everything in the Bible must be true from the very fact that that is what God has said. And assuming that we correctly understand the meaning of the words, we then must believe whatever is said there as a matter of faith.

With all of this in mind, it is not difficult to see why the new heliocentric astronomy came into conflict with the Bible as customarily interpreted in terms of commonsense meanings. If Copernicanism had been proven to be true, then the scriptural passages which appear to say the contrary would need to be reinterpreted accordingly, as Bellarmine was willing to grant.[19] But that was not the case at hand. Rather the issue was what, if any, action should the Church take, given that heliocentrism, even though unproven, was still a challenge to the authority of the Bible and the Church. What ensued was an exceedingly complex series of events, too involved for me to even try to summarize here,[20] which culminated in the most unfortunate decision of the Holy Office to condemn heliocentrism as "false and completely contrary to the divine Scriptures." This decision was published in a decree issued by the Congregation of the Index, dated 5 March 1616. The relevant portion of that decree reads as follows:

> It has come to the attention of this Sacred Congregation that the Pythagorean doctrine of the mobility of the earth and the immobility of the sun, which is false and completely contrary to the divine Scriptures, and which is taught by Nicholas Copernicus in his *De revolutionibus orbium coelestium* and by Diego de Zuñiga in his *Commentary on Job*, is now being divulged and accepted by many. This can be seen from the letter published by a Carmelite priest, entitled *Letter of Fr. Paolo Antonio Foscarini on the Opinion of the Pythagoreans and of Copernicus on the Mobility of the Earth and the Stability of the Sun and on the New Pythagorean System of the World*, Naples: Lazzaro Scoriggio, 1615. In this letter the

said Father tries to show that the above-mentioned doctrine of the immobility of the sun in the center of the
world and of the mobility of the earth is both in agreement with the truth and is not contrary to Sacred Scripture. Therefore, lest this opinion spread further and
endanger Catholic truth, it is ordered that the said Nicholas Copernicus' *De revolutionibus orbium* and Diego de
Zuñiga's *Commentary on Job* are suspended until corrected; also that the book of the Carmelite Father Paolo
Antonio Foscarini is completely prohibited and condemned; and also that all other books teaching the same
thing are prohibited, as the present Decree prohibits, condemns, and suspends them all respectively.[21]

A few days earlier Galileo had been informed of this
decision by Cardinal Bellarmine and was placed under an
injunction (whose specific terms are not clear even to this
day) to obey the decree. Cardinal Boniface Caetani, to
whom Campanella later dedicated the *Defense*, was appointed to undertake the corrections of Copernicus' book
mentioned near the end of the Decree. While all this was
going on, Campanella was writing his *A Defense of Galileo*
in a dungeon in Naples.

After the decree had been published, all discussion of
the merits of the case stopped. The decree called for en-
forcement, not debate. The next sixteen years of Galileo's
life were relatively quiet. The new pope, Urban VIII, who
was elected in 1623, was much more open-minded and in
private conversations encouraged Galileo to continue his
work. This resulted in the publication in 1632 of his master-
piece, *Dialogue Concerning the Two Chief World Systems*,
which ostensively merely gave the arguments on both sides,
but which in fact clearly favored heliocentrism. But tragi-
cally this in turn raised the question of whether Galileo
had violated the ambiguous injunction placed on him in
1616 (which had been unknown to Urban VIII). Galileo's
famous trial and abjuration took place in the spring of

1633.[22] It was concerned totally with the issue of his obedience to that injunction and not with the scientific merits of Copernicanism.

Meanwhile Campanella, who had finally been released from prison in 1629, was forced to watch from the sidelines as the man he had tried to protect in the *Defense* followed in some of his own footsteps. His original question of "whether the philosophical view advocated by Galileo is in agreement with, or is opposed to, the Sacred Scriptures," had finally been answered. But it was not the answer he had proposed.

WHEN WAS THE *DEFENSE* WRITTEN?

Campanella's original autograph or handwritten copy of the *Defense* has apparently been lost. Attempts by several scholars to locate it in various archival collections have not been successful. As a result, the 1622 first printed edition by Tobias Adami is the foundational source to be used in studying the *Defense*. Although that edition is basically reliable, this situation raises some problems about the treatise. For example, from remarks in some of his letters it seems that Campanella's original title was *Apologeticus pro Galileo*, which Adami changed to the stronger *Apologia pro Galileo*, by which the treatise has henceforth always been known but which implies that Campanella's purpose was to "defend" Galileo. However the format of the treatise consists of presenting and evaluating the arguments on both sides, although it is indeed clear that Campanella favors Galileo in his *Defense*.

A much more serious problem relates to the question of when the *Defense* was written. Unfortunately, and for reasons unknown, Adami's first edition assigns no date to the treatise, nor does he give any hints about its date. At first this may seem a minor point, but on second thought it becomes clear that whether it was written before or after 5

March 1616, the date of the condemnation of Copernicanism, makes a considerable difference on how one should read the *Defense*.

If the treatise was composed before the condemnation, then it was written to serve as one of the major briefs to be submitted to the court's consideration in its judgment of the case of heliocentrism. Its "pro and con" style suggests this. More significantly, the work was dedicated to Cardinal Boniface Caetani as having been written at his request, or actually more strongly, "written by your order" (*iussu tuo elaboratam*). Thus Campanella would have been playing a cooperative role in advising the cardinal and the Church on this important issue.

On the other hand, if the *Defense* was written after 5 March 1616, then it takes on an entirely different demeanor. For if the issue had already been settled, then further discussion of the "pros and cons" of the case, especially one that favored Galileo, would have been not only irrelevant but also most unwelcome. In this scenario Campanella would have undertaken an extremely daring and courageous, if not foolhardy, attempt to persuade the Church to change its judgment or to protect Galileo personally from the Church. And this would have been done with an almost reckless disregard for his own personal well-being, since at the time Campanella had long been in prison in Naples and had regularly pleaded for his own release. As we shall see, the same question of dating before or after 5 March 1616 can be, and has been, raised about Campanella's letter of dedication to Cardinal Caetani.

Modern scholars have been divided on this issue. Luigi Firpo's reconstruction of events is that the *Defense* was written in the summer of 1616.[23] His main reason for this date seems to be that there are no documents which refer to the *Defense* before September of that year. Early that month Campanella's friend Pietro Iacomo Failla wrote to Galileo to inform him that a copy of the *Defense* had been sent indirectly to Cardinal Caetani, and that if Galileo had already seen it, Campanella looked forward to receiving Galileo's opinion about its merits.[24] Galileo never replied, despite several further pleas from Campanella in the following months for his assessment. According to Firpo, Campanella became increasingly uneasy as time passed over how the *Defense* would be received. So he decided to take the following daring action, as Firpo interprets the situation:

> Aware that he had done something rash, Campanella did not hesitate to protect himself with a weak defense. Without having even started the project of making the corrections [of Copernicus' book], Cardinal Caetani died in June of 1617. From that moment Campanella became concerned to make it known that the *Defense* had been written in response to an explicit invitation from Cardinal Caetani (which is very improbable), and that it was written prior to the formal condemnation (which is quite impossible, since Caetani was charged with undertaking the revisions ten days after the Inquisition's decree of condemnation.) As a result he added to the text a posthumous dedication to Cardinal Caetani in which he presented the book to him as "written at your request." He then turned it over to his Lutheran friend Tobias Adami, who published it in Frankfurt in 1622. When the first copies arrived in Rome, the bookstores were immediately forbidden to sell them, as Virginio Cesarini informed Galileo on 12 January 1623. A few years later when he was imprisoned at the Holy Office in Rome to give an account of some of his published and unpublished writings, Campanella could not come up with a better excuse than to say that he had written the book prior to the condemnation, and that it was Cardinal Caetani himself who had published it in Germany. This is intrinsically incredible, chronologically impossible, and clearly contradicted by the Preface to the Reader found at the front of the book. This Preface was certainly written by Adami, and he rejects with some

harshness the interference of both Catholic and Protestant theologians in questions of pure science.[25]

Cardinal Caetani died on 29 June 1617. So according to Firpo's interpretation, Campanella added his misleading letter of dedication to the cardinal sometime after that date, and thus about one whole year after writing the *Defense*, in order to create some protection for himself. But that would have been risky, since his self-serving motives could have been revealed by the late cardinal's associates, who might have found a copy of the *Defense* without the dedication among his papers.

Firpo's case relies heavily on the second sentence of Campanella's letter of dedication: "It is up to you to determine what has been correctly said and what should be defended or rejected, since this role has been entrusted to you by the Sacred Senate."[26] This sentence is ambiguous. Firpo takes it to refer to the fact that after the condemnation Cardinal Caetani was assigned to make the corrections of Copernicus' book mentioned in the decree of 5 March 1616.[27] Hence the letter could not have been written before the condemnation. On the other hand the sentence in question may be taken to refer rather to Cardinal Caetani's role in the Church's deliberations on Copernicanism in the weeks prior to the condemnation. In the latter case Campanella could have written the dedicatory letter prior to 5 March 1616.

Further it is difficult to say how probable or improbable it might have been for Cardinal Caetani to have asked Campanella to write a theological evaluation of the orthodoxy of Copernicanism. Caetani was a powerful cardinal, well-educated, open-minded, and widely respected. He was a personal friend of the reigning Pope Paul V and had advised against the condemnation of heliocentrism. His position was quite secure. Would he have turned away from requesting the opinion of Campanella, who, despite his recognized scholarly powers, was after all a convicted heretic residing in prison? On the other hand it is indeed hardly credible that Caetani would have taken steps to publish the *Defense* in Germany at any time after the condemnation, as Campanella later claimed. Also while Firpo correctly refers to Virginio Cesarini's letter to Galileo of 12 January 1623 to establish the fact that sales of the *Defense* were forbidden in Rome, he fails to mention that in that same letter[28] Cesarini explicitly states that the treatise was written before the condemnation of 5 March 1616.

In opposition to Firpo, Salvatore Femiano has argued that Campanella wrote both the *Defense* and its letter of dedication prior to the condemnation.[29] His primary and most forceful argument is that there is not a single passage in the *Defense* itself which either directly or indirectly indicates that it was written afterwards. Rather the whole tone of the treatise, along with some passages which Femiano explictly cites, give the impression that the decision on heliocentrism was still located in the future and thus that the treatise was composed before 5 March 1616. To give just one example, on the last page Campanella indirectly but forcefully advises Cardinal Bellarmine against condemning Galileo's work. The same dating can be attributed to the dedicatory letter to Cardinal Caetani if, as indicated above, its second sentence is interpreted as referring to his participation in the deliberations before any decision was made. For Femiano if the *Defense* had been written afterwards, it would not only have been useless but also would have lost its significance.

Femiano also quotes several comments from Campanella's later writings which support this reading. For example, on 10 June 1628 Campanella wrote a letter to Pope Urban VIII in which he said, "I declare that I wrote the *Apologeticus pro Copernico et Galilaeo* at the request of Cardinal Boniface Caetani, when it was being debated at the Holy Office whether or not their opinion was heretical."[30] Even more significant is a remark made at the end of Article 4, Question X, of Campanella's *Questiones physiologicae*,[31] where he says, "Five years after I wrote this article, I learned that the Fathers in Rome had condemned the theory of the daily rotation of the earth as contrary to the Scriptures, *on the eighth day before our treatise on this topic was received by Cardinal B. Caetani*" [italics added]. According to this comment the *Defense* arrived ironically too late to influence the fateful announcement of March 5th. But also it then clearly would have been written before the condemnation, just as Virginio Cesarini said in his letter to Galileo the next January. Its late arrival would explain why it had no impact on the decision or on subsequent developments, and perhaps also why it has been largely neglected, both then and now, by those interested in the Galileo affair.

Although this matter of dating has not been definitively settled, it seems to us that the textual evidence and arguments justify Femiano's interpretation: "The conclusion is that indeed the *Defense* was written at the request of Cardinal Caetani in February of 1616, and that Campanella sent it to him at the end of that month or at the beginning of the next month."[32]

THE ARGUMENT OF THE *DEFENSE*

The first point to be noted about the Defense is that it was intended to be a theological, and not a scientific, treatise. And hence it should be read accordingly, as Campanella says explicitly in chapter 5.[33] As a result it was not his intention in this treatise to give a direct evaluation on their own merits of the evidence and arguments used in his day for and against heliocentrism. That project he had already undertaken in other writings. So when the reader does encounter discussions of scientific topics in the *Defense*, one should always read them as occurring within a theological context; namely, whether such scientific views agree or disagree with the Scriptures, the early Church Fathers, and the traditional teachings of theology.

The reason behind this, in turn, is that Campanella was a strong advocate of what has come to be called the doctrine of the unity of truth. In short this is the old traditional view that religious truths cannot contradict, or be in disagreement with, the truths discovered by natural reason in any area, including science. His usual way of expressing this view is to employ the common metaphor of God's double gift of truth, i.e., in the book of Scripture and in the book of nature, or God as revealer and God as creator. Since the same good and truthful God is the author of both books for Campanella, they cannot disagree with each other, assuming that each book is properly understood.

If this be granted, then there is an antecedent guarantee that religion and science will not disagree in the long run. The problem rather is to understand each book properly. So Campanella is confident from the start that his initial question, "Whether the philosophical view advocated by Galileo is in agreement with, or is opposed to, the Sacred Scriptures," will be answered in favor of the former alternative, *if* Copernicanism is true. His concern is to show how that agreement would be grounded.

Since the *Defense* is primarily a theological treatise, its method of argumentation is also theological. This means, among other things, that a very prominent place is assigned to proof by an appeal to authority. The main authorities for Campanella are the Bible, the tradition of the Church Fathers, the documents adopted at Church councils, and prominent theologians of the past. The purpose of these appeals, which may seem excessive to the modern reader, is to determine the accepted meaning and the orthodoxy of various points under dispute. Also at times the purpose is the show that the authorities do not agree on a given point, and thus the matter is still an open question. In either case the larger the number of authorities quoted, the stronger is the argument using this method. Such argumentation from authority had had a very long history before Campanella used it. Further, this method was reemphasized by the decree of the Fourth Session of the Council of Trent when it insisted that the acceptable interpretation of Scripture was the one given by the Church and found in the unanimous agreement of the Fathers.

The overall plan of the *Defense* is quite simple. The first two chapters give arguments respectively in opposition to and in favor of Galileo's heliocentrism. The last two chapters answer and evaluate these arguments one by one. The critical chapter 3 in between lays down a series of background assumptions, or hypotheses as Campanella calls them, supposedly needed to formulate the replies.[34]

This format is designed to create an atmosphere of an objective assessment of the arguments on both sides of the dispute, while at the same time Campanella does not at all try to hide the fact that his sympathies are with Galileo. It is important to recall that at that time Campanella was not personally persuaded of the truth of Copernican heliocentrism, although he did think that it was becoming increasingly more probable, primarily because of Galileo's discoveries with the telescope. But he was very strongly convinced that Galileo's freedom to discuss and investigate the matter further should not be blocked, no matter what the fate of Copernicanism would turn out to be.

There is nothing exceptional about the arguments pro and con in the first two chapters. They were common coinage at the time. In general the arguments are based on specific biblical passages, on the historical facts surrounding the publication of Copernicus' book, and on Aristotelian and later Scholastic teachings which were then in vogue. Campanella shows that he was well-informed, since the list of biblical passages quoted in opposition were among the standard scriptural objections of the day.

In the replies to these arguments there are several noteworthy themes in addition to the detailed interpretive discussions of the biblical passages quoted against Galileo.

First, Campanella's lifelong antagonism toward Aristotelian philosophy, which was based on grounds other than the present dispute, comes out strongly in his criticism of the geocentric model of the universe attributable to Aristotle. Second, and related to this, is his harsh criticism of theologians who think that Christian theology, revelation, or religion is tied in any essential way to Aristotelian philosophy. For Campanella this reverence for Aristotle, who held such heretical views as the eternity of the world, the absence of divine providence, and the mortality of the human soul, was an abomination. Rather Christian philosophy should be grounded in Christian sources.

Third, he argues at very great length in chapter 4 that the scattered remarks about cosmology that can be found in the Bible and in the writings of the early Fathers of the Church agree much more with Galileo's than with Aristotle's view of the universe. For instance, according to Genesis there must be water in the heavens since "the firmament divided the waters." But according to Aristotle there cannot be any elemental water in the heavens, which are composed of a distinctively unique kind of matter, the quintessence.

Fourth, at the end of the last chapter Campanella gives the unusual argument that the ancient Pythagoreans of his native Calabria derived their sun-centered astronomy from Jewish sources in the Holy Land. Thus he argues for a continuity between heliocentrism and the historical origins of Christianity.

The main focus of interest for the modern reader falls, however, on chapter 3, which serves as a bridge between the arguments and the replies. One would expect that this chapter would be devoted to methodological points, e.g., the principles of biblical exegesis and the procedures of scientific proof, which would be needed to reply to the objections. But Campanella surprises us by raising a quite different type of issue. Although he had apparently been asked by Cardinal Caetani to write a theological assessment of heliocentrism vis-à-vis Scripture, in chapter 3 Campanella steps back from that topic and asks, "What qualifications must be possessed by anyone who would serve as a competent judge of the original question?"

This was an exceedingly adroit maneuver on Campanella's part. He was not asked, nor did he wish, to judge whether Copernicanism was true or false. That was beyond his competence, and personally he was not firmly persuaded on that issue. But he was competent, and convinced, on the issue of how one should go about dealing with Cardinal Caetani's original question. By turning the discussion in the direction of the qualifications of a good judge, he in effect transcended the specifics of the Galileo case and raised the more general question of the freedom of human thought and inquiry. It is this feature of the *Defense* which makes it a document of perennial interest in human history.

It is perhaps not at all accidental that the man who raised this question was himself at the time a long-term prisoner of the Inquisition. He had experienced the suppression of thought at first hand. And those who later read the *Defense*, whether in Rome or elsewhere, could not have avoided wondering whether the judges of the Holy Office and their advisers in the Galileo case met Campanella's requirements for competence.

Campanella's maneuver is strengthened by the fact that there is little room for disagreement over his list of qualifications for the competent judge of the religious orthodoxy of Copernicanism. Such a judge must be a person who not only loves God but who also is well versed in both science and theology. He needs to know the latest scientific observations and their interpretations in the scientific community; he also needs to know the accepted exegetical principles of biblical interpretation and the history of their employment by theologians. In short, he needs to be a religious believer who is a competent reader of both the book of nature and the book of revelation.

This is what Campanella means when he says that a competent judge "must possess both knowledge and a zeal for God."[35] The ignorant saint and the irreligious scientist need not apply. Furthermore, since the project at hand was the task of assessing the religious orthodoxy of a theory in natural science, the burden of acquiring competence must fall primarily on the theologians who make that assessment.

If this first and overriding condition is met, then the competent judge will thereby know the following: First, the state of astronomy at the time was not yet developed to the point at which the scientific issue could be conclusively settled. Campanella was correct on this point, which counsels caution against a condemnation.[36] Neither the Copernican nor the Aristotelian-Ptolemaic astronomy was proven, and Campanella harshly criticizes advocates of the latter philosophy who thought that their view was beyond criticism. Second, the Bible and the religious tradition do not intend to teach us physics and astronomy, and so it is futile to attempt to settle the scientific issue by quoting biblical or theological authorities. The purpose of religion is to teach us, not about the motions of the heavens, but how to live morally good and holy lives in preparation for the afterlife.

Third, any attempt to forbid Christians to study the book of nature is a crime against Christianity itself. For if the Christian religion is true, then it not only has no fear of other truths, but also should welcome any further knowledge of the natural world as additional insights into the wisdom and goodness of God. In short the Church damages itself if it cuts off any access to God which may be found in the book of nature. Campanella has anticipated C. S. Peirce's later maxim: "Do not block the path of inquiry."

Fourth, theologians should not allow themselves to become so addicted to only one philosophy, e.g., Aristotelianism, that they read Scripture in only one way and thereby lose the richness of meaning in the divine message. This also creates unreal conflicts between the Bible and other philosophical views.

Lastly, not every false statement is a heresy, and so the Church need not be concerned with such errors. The term "heresy" should be applied only to those false statements which directly or indirectly contradict revelation. All other claims should be left to open and unobstructed inquiry, which may in time uncover hidden truths, sometimes where they are least expected. What Campanella is saying here, in effect, is that heliocentrism is not a matter of faith and morals. So the Church should not be concerned about it, even if it thinks it is false. And Galileo should not only be permitted, but should be encouraged, to pursue his "false" view further, for who knows what will come of it.

Such is the perspective of the competent judge of the issue at hand according to Campanella. Were his listed qualifications actually satisfied, either individually or collectively, by the cardinals of the Congregation of the Holy Office and their advisers when they met in February of 1616 to judge the religious orthodoxy of Copernicanism? Most probably not. If so, the advice from Campanella then would have been to suspend action on the matter under these circumstances. This is not bad advice. Inaction would have saved the Church from what became an enormous embarrassment.

Could it be that Campanella thought that he himself satisfied the requirements for the competent judge? We will never know. But if he did, his assessment of the case, which floats through the background of the *Defense*, was both clear and judicious: Do not condemn Copernicanism and do not forbid Galileo and others to discuss the matter and investigate it further. He did not recommend this because he thought that heliocentrism is true, nor because of his great personal admiration for Galileo, nor even primarily because of a commitment on his part to an abstract notion of the freedom of thought. Rather his overriding reason was that such a condemnation would be a serious abuse of power and authority which would become self-destructive for the Church if heliocentrism turned out in time to be true. And he was right. He acted out of loyalty, his fundamental concern being the welfare of the Church which had treated him so harshly. And so he concludes the *Defense* by saying:

> In my judgment . . . it is not possible to prohibit Galileo's investigations and to suppress his writings without causing either damaging mockery of the Scriptures, or a strong suspicion that we reject the Scriptures along with the heretics, or the impression that we detest great minds (especially since in our day the heretics disagree with everything said by Roman theologians, as Bellarmine has pointed out). It is also my judgment that such a condemnation would cause our enemies to embrace and to honor this view more avidly.

POSTSCRIPT: TWO OTHER DEFENSES

One year before Campanella wrote his *Defense*, two other figures in the Galileo case took up their pens to defend the reconciliation of heliocentrism with the Bible. One of them was Galileo himself. The other was the previously obscure Carmelite priest Paolo Antonio Foscarini. Their writings on the topic played a direct role in the condemnation of Copernicanism. However Campanella apparently was not able to read their contributions before he wrote his *Defense*, which shows no influences from these documents composed in the previous year or so.

Galileo was drawn into the dispute after his close friend and colleague, a Benedictine priest named Benedetto Castelli, had informed him that the question of the scriptural orthodoxy of heliocentrism had been raised at the Tuscan Court in an after-dinner conversation. Fearing damage to his own good relations with his royal patrons,[37] Galileo wrote an account of his views on the matter in what has come to be called the "Letter to Castelli" (21 December 1613), which was widely circulated privately. Later in the summer of 1615 he considerably expanded this into the "Letter to the Grand Duchess Christina,"[38] which in time came to be recognized as a work of major importance on the principles of biblical exegesis. This latter treatise was published in northern Europe in 1636, after Galileo's trial. Meanwhile, in the spring of 1615 an adulterated version of the "Letter to Castelli" had been denounced as heretical at the Holy Office in Rome by two Dominican priests, Niccolò Lorini and Tommaso Caccini. Although no charges were brought against Galileo personally as a result of this, his name had become intimately associated with the question of the biblical orthodoxy of Copernicanism before that issue came up for judgment the next year.[39]

In briefest summary Galileo's substantive views in the two letters mentioned above fall into two categories: the topic of the relation between scientific truth and religious truth, and the question of what principles of exegesis should be used in interpreting the Bible. On the former point he argues, following the classic view of Augustine, that if a scientific claim has been proven conclusively to be true, then the Scriptures should be interpreted accordingly. On the other hand, in cases of overlap where it is impossible in principle to establish conclusive scientific proof, one should assent to the biblical account as true. Unfortunately this leaves an unexamined third area of potential conflict; namely, claims in science which are not proven to date but which may become conclusively proven in the future. Galileo was not clear about the status of this third area of overlap, which of course contained the hypothesis of heliocentrism. As a result his attempts to reconcile science and the Bible on the basis of their respective truth claims was not successful.

On the topic of the principles of biblical exegesis, however, Galileo was surprisingly perceptive. The book of nature and the book of revelation cannot be in conflict, for the same reason given by Campanella and many others. However one must realize that the book of revelation is written in the language and in the rhetorical style of commonsense understanding and thus reflects the cultural trappings of the time in which it was written. As a result the Bible speaks in the commonsense idiom of geocentrism, as we all still do today, and as did the Fathers of the Church, who had no reason at all to consider the matter differently. This exegetical principle of cultural accommodation, which was suggested by Galileo and which has been accepted by the Catholic Church since the end of the nineteenth century, had the power to overcome the biblical literalism of Galileo's day. With its use it is quite easy to reconcile commonsense geocentrism with scientific heliocentrism.

Foscarini's approach to the issue was quite different from the path followed by either Galileo or Campanella. He was able to foresee the high likelihood that heliocentrism would be proven true in the near future. In that case, to avoid a conflict between science and the Bible, it would be necessary for the Church to reinterpret those passages of Scripture which would seem to clash with the newly established truth. So Foscarini attempted to perform this task in advance and thus to provide the Church with the needed reinterpretations, just in case. As a result, in January of 1615 he published his *Letter on the Motion of the Earth . . . ,*[40] which identifies potentially conflicting biblical passages and reinterprets them according to heliocentrism. His small book was an exemplary account of how to carry out this theological task.

Needless to say, Foscarini's treatise was not well received in Rome. As we have seen, the Fourth Session of the Council of Trent had reserved the power to interpret Scripture to the authorities within the Church. In this case they thought otherwise than Foscarini about the wisdom of preparing a reinterpretation. Within two months Foscarini's book was censored by the Holy Office. Foscarini wrote a "Defense" of his treatise and sent it, along with a copy of the book itself, to Cardinal Bellarmine for his assessment.

This occasioned Bellarmine's critically important "Letter to Forscarini" (12 April 1615), which crystallized the Church's position into the following three points. First, it is permissible to advocate Copernicanism but only hypothetically and not realistically, since it is contrary to the Scriptures. Second, this is a matter which falls under "faith and morals" as mentioned by the Council of Trent, Session 4, because everything in the Bible is a matter of faith simply from the fact that it is said there. Third, there is no proof to date of Copernicanism. If such a proof were to be given in the future, then at that time it will be necessary either to reinterpret the Scriptures or at least to say that we do not understand the relevant passages.

With Bellarmine's letter the matter was effectively closed. Less than one year later Copernicanism was condemned as "false and completely contrary to the divine Scriptures" in the decree of 5 March 1616. In that document Foscarini's book was also condemned, without any recommendation that it could be corrected, and the book was rigorously suppressed.

In summary, when Copernicanism was condemned by the Catholic Church, there were three defenders against that decision, each with a different emphasis: Galileo, who focused on the principles of biblical exegesis and on the independence of scientific truth from religious truth; Foscarini, who tried to supply an actual reinterpretation of the problematic biblical passages, if that were ever needed; and Campanella, who concentrated on the qualifications required by a competent judge of the case. They failed. But it cannot be said that good arguments in defense of a reconciliation were unknown in Rome.

Thomas Campanella, O. P.

A Defense of Galileo

GREETINGS TO THE BENEVOLENT READER
FROM THE PUBLISHER

For insignificant creatures like us, who live in this world surrounded on all sides like worms in a cheese,[1] it is no small matter to engage in grave disputes about the structure of the world, such as whether our abode or house, which we call the earth, rotates on high around the sun together with the other globes similar to it, or whether the sun rotates around the earth. We are indeed such small creatures that we are very ignorant of such matters. We are like a mouse in a ship who, when asked by a fellow mouse about the ship being at rest on the sea, would never be able to say whether the ship, their common home, is in motion or whether it remains fixed in one and the same place.

As a result many judge these investigations to be more complex than is commonly thought, especially after so many new things have been detected in the celestial globes by means of the optical instrument which the Lyncean philosophers in Rome call a telescope.[2] On the other hand, although certain arrogant people, who wish to pass themselves off as philosophers, have seen most of these things, still their spontaneous amazement did not prevent them from turning others away from a more careful investigation of the truth. And indeed many theologians, both Catholic and Protestant, are especially eager to suppress this investigation by appealing to the unchanging authority of the Sacred Scriptures. But whoever loves the truth must give special consideration to what is right or wrong in this matter. Both in our day and in times past, many famous people who were well informed about both profane and sacred studies, beginning with the Pythagoreans,[3] have defended and still defend this view; and they ought not to be accused rashly of either impiety or ignorance.

These questions and many others are examined in depth by the well-known Italian philosopher, theologian, and monk, Thomas Campanella, in the present treatise, which we wish to submit, dear reader, to your consideration. And lest you think that he is the only Italian cleric to hold this view, you could also read the remarkable and lengthy letter of the Carmelite Paolo Antonio Foscarini,[4] who examines the opinion of the Pythagoreans and of Copernicus on the motion of the earth and the stability of the sun and the new Pythagorean system of the world, written in Italian and addressed to Sebastiano Fantoni,[5] General of the Carmelite Order, and published in Naples in 1615 by Lazzaro Scoriggio. If there were a Latin version of this letter, it would have been inserted as an appendix in this book.

After you have considered and carefully examined the arguments of these writers, and if you also have gone on to read the Germans Nicholas Cardinal Cusa,[6] Nicholas Copernicus,[7] George Joachim Rheticus,[8] Michael Maestlin,[9] and David Origanus;[10] the Italians Giordano Bruno[11] of Nola, Francesco Patrizi,[12] Galileo Galilei,[13] and Redento Baranzano;[14] the Englishmen William Gilbert[15] and Nicholas Hill;[16] and our now famous compatriot Johannes Kepler[17] (who was bold enough to assert along with another writer that, after Galileo's *Sidereus nuncius*, most philosophers are by now Copernicans, as we also conclude), then we have no doubt, kind reader, that you will be a more impartial judge of this very important theory.

Farewell, and look for more writings by this author soon.

[Tobias Adami[18]]

GREETINGS TO THE MOST ILLUSTRIOUS
AND MOST REVEREND
LORD CARDINAL BONIFACE CAETANI,[19]
THE MOST RESPECTED PROTECTOR
OF ITALIAN HONOR,
FROM FR. THOMAS CAMPANELLA

I hereby send to you, Reverend Lord, this treatise, composed at your request,[20] in which the motion of the earth, the stability of the celestial sphere, and the rationale behind the Copernican system are examined in the light of the Sacred Scriptures. It is up to you to determine what has been correctly said and what should be defended or rejected, since this role has been entrusted to you by the Sacred Senate.[21] I submit my opinions not only to the Holy Church, but to anyone who knows more than I, and especially to you, the protector of the muses of Italy, who will not die as long as you live. Therefore may you live forever. Amen.

A DEFENSE OF GALILEO

Introduction

Some time ago I answered two of the most pressing questions of our times;[22] namely, whether it is allowable to construct a new philosophy; and whether it is allowable and expedient to oppose the Aristotelian sect and the authority of the pagan philosophers and to replace them in the Christian schools with a new philosophy which follows the teachings of the saints. I am now invited to respond to another specific controversy raised by those who are upset by the philosophical view advocated by Galileo of Florence because that teaching seems to be contrary to the Sacred Scriptures. I will respond as best as I can.

The question before us then is *whether the philosophical view advocated by Galileo is in agreement with, or is opposed to, the Sacred Scriptures.*

I will fully answer this question in five chapters. In the first I will give the arguments against Galileo; in the second, the arguments in his defense; in the third I will establish some hypotheses which are needed for the twofold resolution which follows; in the fourth I will respond to the arguments against Galileo; and in the fifth I will explain how the arguments in his defense are to be evaluated.

CHAPTER 1

The Arguments
against Galileo

The first argument against Galileo is that it seems that theological doctrines would be completely overthrown by anyone who tries to introduce new ideas which are contrary to the physics and metaphysics of Aristotle, on which St. Thomas and all the Scholastics based their theological writings.

2. Further, Galileo publishes opinions which contradict all the Fathers[23] and the Scholastics. For he teaches that the earth moves and is not in the center of the world, and that the sun and the stellar sphere are at rest. But the Fathers, the Scholastics, and our senses testify to the contrary.

3. Further, he clearly contradicts Sacred Scripture. For it is said in Psalm 92 [92:1], "He established the orb of the earth, which will not move," and in Psalm 103 [103:5], "He established the earth above its foundation; it will not move from age to age." And Solomon says in Ecclesiastes 1 [1:4], "But the earth stands forever."

4. Further, the same thing is clear regarding the motion of the sun. For in the same place in Ecclesiastes [1:4–6] it is said, "The earth stands forever; and the sun rises and sets and returns to its own place; there it rises and rotates through the south and turns toward the north; the wind crosses through the universe in a circle and returns in its circles."

5. Further, in Joshua 10 [10:12–13] there is related the most astonishing miracle that Joshua stopped the motion of the sun with his words, when he said, "Sun, stand still over Gibeon, and moon, over the valley of Aijalon. And the sun stopped in the middle of the heavens and did not continue its motion for a period of one day." The same account is found in Ecclesiasticus 46 [46:5].

6. Further, in Isaiah 38 [38:8] God gave to Hezekiah, as a sign that he would recover his health, the miracle of Ahaz's clock: "The sun turned back ten markers through which it had already crossed." Hezekiah was questioned about this miracle by the king of the Caldeans, who had noticed this retreat of the sun because he was devoted to astronomy, as we know from 2 Paralipomena [2 Chronicles 32:24, 31]. Hence if God had not truly stopped the motion of the sun, these would not be true miracles. Therefore the Scripture would be false, since it relates these two events as true miracles.

7. Further, the Sacred Scriptures speak with admiration of the motion of the starry heavens. For in the song of Deborah in Judges 5 [5:20] it is said, "The stars remain in their order and courses, and fight against Sisera." Therefore the stars move, and hence so do the heavens, in which they are inserted like knots in a board. Also the Apostle Jude [Jude 13] speaks of "errant stars," hence they move. Also in 3 Esdras 4 it is said, "Great is the earth, and excellent the heavens, and the speedy course of the sun crosses the heavens in a circle and returns to its place in one day."[24] Hence when Galileo says that the starry heavens are immobile, he openly contradicts the word of God.

8. Further, Galileo maintains that there is water on the moon and on the planets. But this is false because those bodies have an incorruptible nature, since Aristotle and all the Scholastics testify to the eternity and immutability of the heavens for all ages. He also maintains that there are mountains on the moon,[25] and that there is elemental earth there as well as in the other celestial bodies. But this seems

too much to cheapen the abode of the angels and to destroy the hopes which we have placed in the heavens.

9. Further, from Galileo's opinion it follows that there
are many worlds and earths and seas, as Mohammed said,
and that there are human beings living there, if the four elements of our world also exist in the stars. For if every star
is composed of all four of the elements, then clearly each
will itself also be a world. But this seems to be contrary to
the Scriptures, which speak only of one world and one
human race. I will omit the point that to say that Christ
also died on other stars to save those inhabitants is to re-
vive the heresy, which some have maintained, that at one
time Christ was crucified a second time in the other hemisphere of the earth to save the humans living there as he
has saved our part of the world. It would also be necessary
to agree with the heretic Paracelsus[26] that there are other
humans who live in the air and in the waters and under the
earth, who enjoy beatitude, even though it is doubtful
whether they are included in the redemption. The Jesuit
Martin Delrio has written against this in his *Disquisitiones
magicae.*[27]

10. Further, it does not seem possible to debate these
matters without creating an immense scandal. For the
schools already have an established doctrine about the
heavens and the earth, and it agrees with theology, as
the Scholastics teach. Hence those who teach anything else
seem guilty of trying to create a new way to destroy Scholastic theology and of pridefully placing themselves above
others.

11. Further, the Scriptures warn us, "Do not inquire
about the higher things" [Ecclesiasticus 3:22], and "Do not
wish to know more than one needs to know" [Romans
12:3], and "Do not go beyond the limits set by your fathers" [Proverbs 22:28], and "He who seeks grandeur will
be oppressed by ambition" [Proverbs 25:27]. But Galileo
seems to do just the opposite of this, because he subjects
the heavens to his own cleverness, and he makes the archi-

tecture of the whole world conform to his approval. Cato teaches more properly: "Do not search out the secrets of God, or inquire into the heavens; since you are mortal, attend to the things which are mortal."[28]

CHAPTER 2

The Arguments
for Galileo

On the other hand, in Galileo's favor there is the authority of the theologians who permitted the publication of Nicholas Copernicus' book on the revolution of the spheres, based on his observations beginning in 1525, because it did not contain anything contrary to the Catholic faith. In that book he advocated that the earth moves, that the firmament, i.e., the starry heavens, are immobile, and that the sun is at rest in the center of our world. But Galileo does not announce anything new except other, previously unknown, bodies.[29] Therefore if Copernicus' book does not disagree with the Catholic faith, then neither does Galileo.

2. Also, Copernicus' book was approved by Pope Paul III Farnese,[30] to whom the book was dedicated, and by certain cardinals (who examined the manuscript before it was published, as is obvious from the introductory letters).[31] In the days of Paul III men of the most outstanding genius flourished within the Church, and they were assembled from all sides and supported and favored with honors by that pope, who was most noble in soul, in virtue, and in blood. Hence it would be astonishing if those men were as blind as moles in regard to Copernicus while our contemporaries, who are of no great reputation, are more ob-

servant from afar than Argus[32] and have made more certain observations than Galileo.

3. Also, after Copernicus' time Erasmus Reinhold,[33] J. Stadius,[34] Michael Maestlin, Christopher Rothmann,[35] and many others defended the same opinion. These more recent astronomers found it impossible to establish astronomical tables correctly without using Copernicus' calculations and to explain celestial motions properly without using the most certain mathematical principles, the evidence of the senses and of all peoples, and the theories of Copernicus. And this is not true only of recent astronomers. Even before this, Francesco Maria of Ferrara[36] maintained that, because of observations of new appearances in the heavens, it was necessary to formulate a new astronomy, which his disciple Copernicus provided.

4. Also, the most learned Cardinal Cusa has accepted this view and has acknowledged other suns and other planets rotating in the starry firmament. This opinion was also defended by a person from Nola[37] and by others who, being heretics, we cannot mention by name. But they were not condemned as heretics for this reason; nor have any Catholics ever been prohibited from publishing on this topic. Other defenders are the illustrious Johannes Kepler, the Imperial Mathematician, who supports this view in his treatise[38] on Galileo's *Sidereus nuncius*, and the most ingenious Englishman William Gilbert in his book[39] on magnetic philosophy, as well as numerous other Englishmen whom I will not mention.[40] Another defender of this view is Giovanni Antonio Magini, a mathematician in Padua, who announced in his astronomical tables[41] from 1581 up to the present year of 1616 that he used the calculations of Copernicus and Reinhold, and who argued against those who think otherwise.

5. Also, the Jesuit Father Clavius,[42] in the last edition of his writings, advises astronomers to work out a new system of the heavens, since he had become convinced that Mercury and Venus rotate around the sun, even though the

followers of Aristotle had previously thought otherwise. Observing this warning, a contemporary mathematician, the so-called Apelles,[43] in his observations of sunspots, leans toward the view of Galileo and Copernicus.

6. Furthermore we will show in our conclusion how very old is Galileo's opinion regarding the motion of the earth, the stability of the sun in the center, and the composition of the stars, including their containing water and the other elements. Indeed this view originated from Moses himself. Pythagoras, who was of Jewish stock although he was born in a Greek city, as St. Ambrose[44] testifies, brought this idea to Greece and to Italy, and taught it at Croton in Calabria. Aristotle attacked this theory with empty arguments and without mathematical proof, using instead moral and boorish conjectures, just as he spurned the books of Moses because with his logic he was unable to grasp their depth and mystery and profound explanations, as we know from the writings of St. Ambrose and Pico della Mirandola.[45] But Galileo has saved our ancestors from the damage caused by the Greeks. This same view was held by Numa Pompilius,[46] a student of Pythagoras and the wisest king of the Romans, as not only Ovid[47] but many historians testify, although others deny it. Pliny rightly says[48] that Pythagoras was the wisest of philosophers by decree of the Roman senate when they dedicated a statue to him (after having been ordered by the Delphic oracle to erect and dedicate a statue to the wisest of the Greeks.) Therefore those who complain about Galileo's philosophical views and teachings, and who prefer Aristotle to the Pythagoreans, seem to cast an insult against Italy, against Moses, and against Rome, now that the truth that has been buried has come to light. Our ancestors, however, are not to be blamed, for they had not yet discovered the new earths, and celestial worlds, and new phenomena, and the agreement of the Scriptures with this philosophy.

7. Since theologians, from the time of Casella[49] and Francesco Maria of Ferrara[50] up to the present day, have

not only not condemned this astronomy but have ordered its publication, and since the number of such theologians has not decreased in more recent times, it would seem that the opponents of Galileo are not motivated by zeal for the teaching of Christ but rather by either envy or ignorance.

8. Also, in the Scriptures the starry heavens are called the "firmament" because they are immobile. Therefore the earth moves. And hence the sun is at the center. For in this way all the phenomena and the principles of the mathematicians are saved, as Copernicus and his followers have proven. Even the followers of Ptolemy admit this.

9. Also, sunspots and new stars in the starry heavens and comets above the moon clearly show that the stars are other world systems.

10. Also, we will prove below from the most holy Doctors of the Church that the text of Moses cannot be correctly explained unless the stars are other worlds.

11. Also, St. Justin,[51] in his *Questiones ad orthodoxos*, teaches that the Christians and the pagans disagree about the shape of the heavens; the latter saying that they are spherical and mobile, the former saying that they are like a vault and are immobile. And other doctors say that the heavens are called a "firmament" because they are immobile.[52]

CHAPTER 3

Three Hypotheses
Are Established for
the Twofold Resolution
Which Is to Follow

efore I respond to each set of arguments given for
and against Galileo by the ancient and the modern
theologians, I will first establish here some reliable
foundations or hypotheses which are fully consistent
with the teachings of the saints, with the laws of
nature, and with the views of everyone.

FIRST HYPOTHESIS

Anyone who wishes to become a judge of an issue
relating either in whole or in part to religion must possess
both knowledge and a zeal for God, as St. Bernard teaches
in his *Apologia*[53] in agreement with the words of the
Apostle in Romans 10 [10:2].

The first part of this double claim is proven as follows. Those who have knowledge, but no zeal for God,
cringe before those who govern the tribunals or schools,
and thus do not dare to speak out for the truth, as is said
in John 12 [12:42–43], "Many of the leaders believed in

Jesus, but because of the Pharisees did not acknowledge it lest they be expelled from the synagogue; they held human glory in higher esteem than the glory of God." Also, in Romans 1 [1:21–23] the Apostle condemns the philosophers for not paying homage to God even though they knew Him. Rather they made sacrifices to false gods because they were fearful of being accused in the senate of criminal heresy, as is related by Plato in his *Apology* for Socrates, and by Xenophon, Cicero, Pliny, and others. Indeed many of them were put to death as irreligious. But there were others who, in order to appear at least to be concerned with the public welfare, defended the commonly held point of view, and thus gained money and honor; they did not perceive or work for truth and justice, but for a little glory and their bellies. And in abandoning their own judgment, they criminally accepted someone else's, as Pope Leo said of Pilate.[54] As the Apostle said [Romans 1:25], "They have given up the truth of God for a lie," and as a result have so changed themselves that the truth appears to them to be what they defend with their mouths but deny with their hearts. This results in a plague of the mind, as Livy[55] has said and also we ourselves in our *Antimachiavellismus.*[56]

The second part of our double claim is proven as follows. Those who have zeal for God, but not knowledge, cannot judge a religious question, no matter how holy they are, unless they have received a revelation directly from God. Thus in Romans 10 [10:2] the Apostle testifies that the Jews imitated the Christians ". . . in their zeal for God but not in their understanding." He also says that they thought that they had thereby submitted themselves to God. And although he was taught and learned in the law by Gamaliel[57] and in secular studies, still Paul said [1 Timothy 1:13], "I acted out of ignorance and from my lack of belief . . . ," for he had not yet examined all the arguments for the Christian faith, as he should have.

Furthermore, although Lactantius Firmianus[58] and St.

Augustine[59] were holy and learned men, they denied the existence of the antipodes[60] because of their zeal for God and the Scriptures. This is clear from the arguments which they used; namely, that humans living there would not be descendants of Adam, which is contrary to Scripture, because it is impossible to travel from here to there across the impassable ocean. Others add that Christ would have had to be crucified twice, once here and once there, and that the Scripture says that the heavens are spread out like a vault, whose base is the earth (as Justin says), above which is water and above that are the immobile heavens. However we now know that these arguments are false because of their deficiencies in mathematics and cosmography, and hence the Scriptures have also been distorted. And the opinion of St. Thomas that no humans live in the region under the line of the equinox[61] is also seen to be false, both because he was deficient in physical and geographical knowledge and because of his zeal for Aristotle, whom he chose to believe rather than the arguments of Albert the Great[62] and Avicenna.[63] And also, because of zeal for the Scriptures, St. Ephrem,[64] Anastasius Sinaita,[65] and Moses,[66] Bishop of Syria, maintained that there is a terrestrial paradise located in the whole of the other hemisphere of the earth, for without such a very large place, they said, there would not be enough room for the four rivers of paradise and for so many trees and so many other things. Nevertheless we now know from the testimony of navigators that they were wrong.

Hence we rightly conclude that without knowledge even a saint does not judge correctly. Therefore in chapter 11 of his small treatise against those who attack religious orders because of their devotion to philosophy,[67] St. Thomas alludes to this in a gloss on Daniel 1 when he says, "If someone who is ignorant of mathematics attacks the mathematicians, or if someone ignorant of philosophy attacks the philosophers, who would not laugh even if he were ridiculed for his laughter?" And the comic poet says of such a judge, "By the immortal gods, nothing is worse than an ignorant man who decides that nothing is right unless it pleases him."[68]

SECOND HYPOTHESIS

To be able to judge correctly, there are six things which a judge of a religious question must know.

First, a speculative theologian who would enter into a dispute against sectarians must have a philosophical knowledge of both celestial and terrestrial matters.

Second, astronomy has not yet been perfected by the philosophers.

Third, neither the Lord Jesus nor the holy man Moses have revealed physics and astronomy to us, for it is said, "God handed the world over to the disputes of men," Ecclesiastes 1 [1:13], and "The invisible things of God come to be understood through the things which He has made," Romans 2 [1:20]. Rather they have taught us how to live in holiness and to understand supernatural teachings, for which nature is not adequate.

Fourth, anyone who forbids Christians to study philosophy and the sciences also forbids them to be Christians. Also only the Christian covenant recommends all the sciences to its members, because it has no fear of being shown to be false.

Fifth, there are those who would use a dogma of the Christian faith to attack philosophers who have proven their own doctrines with evidence and argumentation, even though those doctrines are not expressly contrary to passages in the Sacred Scriptures whose meanings are not open to a different explanation from other passages. Such a person is dangerously self-destructive, is a threat to the faith, and is open to ridicule from others. Even much worse is the person who would interpret the meaning of Scripture in terms of only one philosopher with the result that other philosophers come into conflict with Scripture.

Sixth, not every false statement is so contrary to the Scriptures that it should be treated as a heresy in the Church Militant, as perhaps it is in the Church Triumphant,[69] as long as it does not deny the meaning of Scripture either directly or in its consequences. And if a theologian defends a doctrine which seems to be more or less contrary to God's Scripture, then no one should be condemned or forbidden to undertake a futher investigation to determine whether the doctrine as presented should be held, as long as he carries out that inquiry with a mind open to the truth and not to the destruction of the faith.

These six assertions have already been proven in our *Theologia*,[70] but it is no trouble to prove them again here in order to deal with the question at hand.

PROOF OF THE FIRST ASSERTION

A Christian needs to know only what must be believed to attain eternal salvation, as all theologians teach, including St. Thomas in his *Summa Theologica*, II–II, 8 and 9.[71] However this is not adequate for the theologian, whose role is "to encourage others with sound doctrines and to refute opponents" [Titus 1:9], as the Apostle and all the Fathers teach. For the theologian needs to judge all things in terms of their ultimate cause, which is God, and not merely in terms of their lesser causes, as do other authors and wise men.

Therefore he needs to be thoroughly acquainted with all the sciences in order to know both God, his principal object, and all the works of God, and in order to be able to argue against and attack any science which might contradict divine science in its treatment of God or of the works of God among men. For one truth does not contradict another truth, nor does an effect contradict its cause.

Hence human science does not contradict divine science, nor do the works of God contradict God, as we have been reminded by the Lateran Council[72] under Leo X. As a result in his small treatise against those who attack religious orders,[73] St. Thomas shows that friars should take up the work of secular knowledge and eloquence, and that anyone is blind who cannot see how necessary and useful the sciences are to the theologian.

Therefore, although theology in itself does not need proofs taken from the human sciences, nevertheless for our sake theology does need to do this so that we can strengthen our convictions by understanding the supernatural in terms of the sensible and the natural. This is proven by the testimony of Augustine, Jerome, Dionysius, and other Fathers, who not only teach that one should do this but also do so themselves. In his letter to Magnus, Jerome[74] says, "One does not know what ought to be more admired in them, secular learning or knowledge of the Scriptures," and he adds that for this reason the Apostle Paul read the poets and the philosophers, whom he also often quotes. And in his *Moralia*, when commenting on Job [9:9], "He made Arcturus and Orion," Gregory[75] says that this was derived from the wisdom of secular astronomers. The same thing is proven by the Fathers and by St. Thomas in [his *Summa Theologica*] I, 1, from the saying of Solomon, "Wisdom (i.e., theology) called for protection from its handmaidens (i.e., the sciences)" [Proverbs 9:3].

Now it is clear that the sciences exist in the human race as a whole, and not only in this or that individual person. For God made man to know God, and by knowing to love Him, and by loving to please Him; and for this man has senses and reason. But if the purpose of reason is to attain knowledge, then humans would act contrary to the divine natural order, just like a man who would not wish to use his feet to walk, unless one uses this gift of God according to the divine plan, as Chrysostom[76] has regularly argued. As Aristotle has said, "All men by nature desire to know,"[77] and as Moses said in Genesis 1 [2:15], "God put man in paradise to cultivate and take care of it." But this was not manual labor or the caring for animals; humans who lived then were born spontaneously and without labor, and all the animals obeyed man. Rather man's work was to know things, and to observe the heavens and the natural world out of curiosity, so that he would as a result investigate everything to meet his obligation to venerate God (which cannot be done without first having knowledge, for "The invisible things of God are known through what He has made," as the Apostle said [Romans 1:20]).

Even if it be granted that all the sciences were infused into Adam, still he lacked experiential knowledge. Further, this command to learn was given to him, not as an individual person but as the head of the human race, and hence it has been also given to us, his descendants, as the Fathers testify. David says the same thing, "Seek God and your soul will live" [Psalm 69:32]. But we can seek God only in the natural things he has created, as one searches for a cause in an effect. Elsewhere it is said, "Your works are wonderful, my soul searches into them" [Psalm 119:129]. And in Ecclesiastes 1 [1:13] Solomon says that he has carefully investigated all things under the sun, even though he had been given infused knowledge. In Wisdom 7 [7:16–21] he reveals further that he knew all things in natural science, mathematics, astronomy, and logic. And in 3 Kings 4 [1 Kings 4:33] it is said that he discussed all things in physics, or as others maintain, that he had written about plants, birds, stones, and fish.

As a result, from the beginning the world has been called the "Wisdom of God" (as was revealed to St. Brigid)[78] and a "Book" in which we can read about all things. Hence in his Sermon 7 on the fast days of the tenth month,[79] St. Leo says, "We understand the meaning of God's will from these very elements of the world, as from the pages of an open book." And in Sermon 8 he proves this with these quotations, "The heavens declare the glory of God, etc.," and "The invisible things of God are known from the things He has made." And indeed as Cyril says in Book I of his *Contra Julianum*,[80] "Philosophy is a catechism for the faith;" whoever despises philosophy opposes the faith. And also in his sermon entitled "I heard what the Lord said to me," St. Bernard says that the world is the book of God, which we ought to read regularly. According to Nicephorus,[81] St. Anthony[82] said the same thing, and so did Chrysostom in commenting on Psalm 147 [147:20], "He never does this for other nations."[83] Hence no one can be excused for not accepting this ordinance, for "Their voice goes out through all the earth" [Psalm 19:4].

Corollary: Since the more wonderful and more extraordinary things in the world are better images of God, their author, they should be investigated for this reason with greater care. And by this study divinity is shown to the human soul. Such are the heavens and the stars and the great system of the world. Thus Anaxagoras has said that man was made to contemplate the heavens. And Ovid was much praised by all the theologians, and especially by Lactantius, for having said of God, "While the other animals look down towards the earth, he created man to face upwards, and he ordered him to see the heavens and to stand erect, turning his gaze to the stars."[84]

David reveals the reason for this when he sings in Psalm 18 [18:1], "The heavens proclaim the glory of God, and the firmament speaks of the work of his hands," and in Psalm 8 [8:3], "For I will look at your heavens, the work of your fingers, and at the moon and the stars, which you have made." Moreover Plato in his *Epinomis* and in his *Axiochus*[85] (assuming that the latter was not written by Xenophon) proves the dignity and the deification of man and the immortality of his soul from his knowledge of the heavens, for example, the stars, the equinoxes, eclipses, etc. We have also said much about this in our *Antimachiavellismus*.[86] Ovid also confirms this when he says to the astronomers, "Yours is a happy lot because your primary

role is to know about these matters and to rise up to the
celestial houses; you bring the distant stars closer to our
eyes, and you subject the heavens to your genius."[87]

These praises belong to Galileo more than to anyone
else, as we have said publicly elsewhere. I leave out here
what Josephus[88] and Philo[89] said about physical and celestial
knowledge; and what Berosus[90] noted about this in Noah
and Abraham; and how the patriarch Jacob used his physical knowledge to escape from Laban's greed and became
wealthy, as the Scriptures testify [Genesis 30:25–43]; and
how the ancient fathers lengthened their lives by using their
knowledge.

Furthermore God produced signs of his first coming
in the heavens and on the earth, as he said in Haggai [2:7],
"In a little while I will move heaven and earth, and the
desired one will come for all people." And we have proven
in our *Articoli profetali*[91] that this has actually happened
by examining the changes of the eccentricities, of the
equinoxes, of the obliquity of the ecliptic, and of the
apogees, which began then and are observable now. Also
Luke 21 [21:25] speaks clearly of the signs of the next
coming in the sun, the moon, and the stars. But the ancient
astronomers attributed these signs to false causes. As the
Apostle Peter prophesied in 2 Peter 3 [3:3–4], "Scornful
men, following their own desires, will say during the last
days," (as do the Aristotelians and the Machiavellians
now), "where is his promise and his coming? From the
time that our fathers have fallen asleep, everything has
remained the same back to the beginning of creation."
Against this I have shown that not everything has
remained the same from the beginning; rather signs have
appeared in the sun and the moon and the stars. In his
commentary on Luke 21, St. Gregory used a physical argument dealing with meteorological changes to prove correctly that these signs occurred close to his own day.[92]

As a result, those who would prevent the observation
and verification of changes in the heavens are like those

other sons of darkness who wish that "the day of the Lord would come to take us like a thief in the night," as St. Paul teaches in 1 Thessalonians 5 [5:2–5], where he warns us to be vigilant lest we become sons of the night. But he who watches is the one who observes the signs displayed in the sun and the moon and the stars, and not those who are like the Jews of old who disregarded the signs in the star of Balaam[93] and engraved their obstinacy in stone, as Augustine tells us. Therefore in the Scriptures, which is the first book of nature, we believe the Apostles above all others, just as David said of them, "Their voice spreads throughout the whole earth, yet they do not speak, etc." [Psalm 18:4–5] And Paul says the same thing about the Apostles in Romans 5 [15:21]. For the two books of God agree with each other.

PROOF OF THE SECOND ASSERTION

No philosopher or theologian has yet spoken with adequate competence or certitude about the nature, order, position, number, motion, or configuration of the heavens or about the construction of the universe. And so it is not yet possible to speak of these matters with precision. This is proven both from the Scriptures and from the diversity of views among the experts. For first it is said in Job 38 [38:33, 37], "Who knows the order of the heavens, and gives their law on the earth?" and a little later, "Who will announce the law of the heavens?" Second, Solomon also says in Ecclesiastes 9 [3:11], "God gave the world over to human investigation, but in such a way that man's efforts will not discover what God has done from the beginning up to the end." In chapter 8 this is repeated, along with many other similar things.

Furthermore, one would have to be insane to think that Aristotle has established the truth about the heavens and that there is nothing more to be investigated. For in

De caelo, II,[94] Aristotle tells us what he has accepted from the Egyptians; namely, there are eight spheres, including the sphere of the fixed stars. This latter is the first sphere, which in a period of twenty-four hours moves all the spheres of the planets with a violent motion from east to west contrary to their own inclination, while by their own natural motion they move more slowly from west to east. And the moon moves twelve degrees of the 360 degrees which are completed due to the violent motion of all the spheres in their daily rotation.

But in *Metaphysics*, XII,[95] he does not wish to say that the first sphere moves all the others. Rather each is moved by its own intelligence; and these intelligences are multiplied in such a way that there are as many intelligences as there are appearances and motions. However he gives no account of these appearances, as St. Thomas, Simplicius,[96] and other commentators have pointed out. Furthermore he establishes a war between God and the angels, since they move contrary to His motion, and though they are said to imitate God, in fact they oppose Him. In the same way there is a war among the angels, since they try to move against each other, one to the east, another to the west, one to the north, another to the south. He maintains that just as many angels move the spheres as retard them, with the result that he introduces not only discord in both the heavens and in the angels, but also disorder and weariness in the generation of motion.

He gives no explanation of why the stars seem to move upwards or downwards at different times, or why sometimes they are stationary or move quickly forwards or slowly backwards, or why there are changes of the eccentricities and apogees and equinoxes. And he cannot explain these matters, since the heavens are composed of the fifth essence. Nor could he explain why Mars was observed much later by Tycho to pass below the sphere of the sun, or how there could be spots on the sun, new stars in the stellar sphere, and comets above the moon.

Therefore Aristotle's astronomy must be completely false, since it does not account for these matters which have been established by the senses and by the most reliable instruments. For this reason St. Basil[97] and St. Ambrose[98] declare that those who say with Aristotle that the heavens are composed of the fifth essence are heretics. And they also deny that the sun is formally hot, as we will show below and as we have proven in our *Questiones*[99] in defense of the philosophy of the saints. We will pass over the fact that he locates the sun immediately above the moon, which is false according to St. Thomas and the followers of Aristotle himself.

Aristotle admits that he knows nothing about the heavens; and he entrusts the job of investigation to those who are more versed in this science, as is clear in *Metaphysics*, XII.[100] He confesses that he has derived his ideas from Calippus and Eudoxus,[101] and has added only the rotation of the spheres, which aggravates the war among the angels. We will also pass over the impieties which result from the fifth essence and from the eternity of celestial motion, for St. Thomas and the Christian commentators have explained this clearly. St. Thomas gives arguments to refute this in Lecture 10,[102] where he teaches that Aristotle held that the eternity of motion is real and not just hypothetical, for otherwise there would be no God. This would make atheists of us who deny the eternity of motion. But St. Thomas refutes this.

I will never be sufficiently astonished at those potbellied theologians who locate the limits of human genius in the writings of Aristotle. The fact that not even Ptolemy reached the truth is shown by the new phenomena which his theory cannot explain, and thus he does not remove disorder from the heavens. I will also pass over the errors which Copernicus introduced into astronomy, for example, that there is a regular motion of a sphere around a center other than its own,[103] and other such things. Thebit[104] and King Alfonso[105] invented librations and new spheres, but Copernicus has shown that they were wrong and has returned to the teachings of the ancient Pythagoreans, which provide a better account of the appearances. In addition to this, Galileo has discovered new planets and new worlds and previously unknown changes in the heavens.

Therefore anyone is insane and most ignorant to think that an adequate knowledge of the heavens is to be found in Aristotle, who contributed nothing on his own and who encouraged others to investigate such matters. And those who came after him are uncertain and are still fighting with each other.

Appendix: But someone might object that if an adequate and true science of astronomy is not possible, as Job says, then it is better to stop rather than to continue this useless inquiry. This is true; however he who does continue this investigation is not thereby a heretic, as is implied, even though it is perhaps useless to procede.

On the other hand, our natural desire always to learn more shows that further inquiry is not useless. For St. Bernard, in chapters 4 and 5 of his *De considerationibus* for Eugene III,[106] says, "Although what God is will not be discovered, nevertheless it is always most worthwhile to make such an inquiry." Now the study of celestial matters is also directed towards God, whom we have been ordered always to seek. And even though we will never perfectly comprehend the God whom we must seek, as Paul says to the Athenians [Acts 17:27], nevertheless we always find something more, and thus we become a little more like God.

It is better (as Aristotle says in *De anima*, I)[107] to know a few things with probability about important matters than to know much with certainty about insignificant matters. Thus after the Caldeans many things were discovered in the heavens by the Egyptians, and later many more by the Greeks, and today still many more by the Germans and the Italians. And it is astonishing how many more vistas, in which God reveals his wisdom, power, and love, have been discovered by Galileo.

Saints Leo, Anthony, Bernard, Chrysostom, and others say that the world is the book of God, which we must study diligently. Thus in one of his sermons Bernard teaches that those who have not received the grace to seek God in supernatural things should seek him in natural things, for we are elevated from the latter to the former. The same thing is clearly proved by Richard of St. Victor[108] in his *Benjamin* in his sermon on contemplation.

Reason also establishes this. For if God created the heavens for his own glory, as Solomon says, then he certainly wishes us to admire and praise them and to honor God as their author; just as a talented painter or poet wishes that his pictures or poems be known, and that his artistic excellence be recognized, and that the artist be praised. In the study of the heavens the divinelike character of the human soul is well displayed and increased, as we have said. Therefore such an investigation is not useless.

There are envious people of little talent and faith in God who think that we should stop studying Aristotle and the other ancient philosophers and not inquire any further, especially after the light of the Gospel and after the discovery of new stars and the New World, which were unknown to the ancients along with the light of faith. This would bring our nature to a perfection superior to the ancients and free us from their yoke, for as Cyril testifies, their philosophy is a catechism, while ours is a perfected doctrine. As a result we would be better able to understand the world, which is the book and wisdom of God, as long as we did not neglect the grace which is within us.

I agree with this, assuming that other conditions are equal. I am not saying that the mind of a rustic Christian is equal to the mind of Plato. But I am saying that if minds are produced which are equal to Plato and others, then after the Gospel they would be more proficient in the sci-

ences than were Plato and the others. Plato himself said in
the *Hippias*[109] that contemporaries are not inferior to the
ancients, except for the envy of the living and the venera-
tion of the dead.

That we should not stop inquiry is proven from the
facts that God is good to those who seek him, as Jeremiah
says,[110] and that it always reveals new things, as we have
seen above. And as St. Bernard said, "As long as you cling
to the same things, you will not find anything else."
Therefore it is not always useless to inquire. As St. Leo
said, "In divine things he who thinks that he has found,
has not found what he seeks but has sought in vain."

PROOF OF THE THIRD ASSERTION

After we have seen that the ancient philosophers did
not set any limits for physics and astronomy, it is easy to
say the same thing about Christ and Moses. For in the
Gospel Christ is never found to discuss physics and astronomy but only morality and the promise of eternal life,
whose pursuit he has made clear by his example, teaching,
and blood. Moreover such discussions would have been
quite superfluous. For in the beginning God gave the
world over to human investigation so that we might know
and worship God through the things which he made.[111] For
us to be able to do this, he gave us a rational mind, and
for avenues of investigation he provided the five senses as
windows to the mind, as the Apostle Peter taught, according to St. Clement.[112] With these as our means we should
look upon the world as the image of God, and admire the
things that are in it, and seek out God its author, as
Chrysostom says in his commentary[113] on Psalm 147 and
in many other places, for original sin did not diminish our
natural powers, as all the theologians agree.

Therefore it would have been superfluous for him,
who came to redeem us from sin, to teach us what we are

able and obliged to learn on our own. As a result he did not order the Apostles to teach such things, but rather to baptize and to teach what he himself had done and taught, as is said in the last chapter of Matthew [28:19–20], and to prove this through miracles and martyrdom, as is said in the last chapter of Mark [16:15–20]. Also, in his *Sermo in festo apostolorum Petri et Pauli* [I, 3], Bernard says, "The Apostles did not teach the art of fishing or of building theater scenes or any other such things; nor did they read Plato or contest the subtleties of Aristotle, etc.; rather they taught me how to live, etc." Furthermore in Book I of St. Clement's *Recognitiones*,[114] when Barnabas[115] was asked by a Roman philosopher why nature has given six legs to the very small mosquito but only four legs to the very large elephant, Barnabas replied that he had been commissioned by Christ to teach about the kingdom of heaven and not about physical things which can be investigated naturally. The Apostles did not argue against philosophy. On the contrary, Christ praised the Pharisees for predicting rain and fair weather from the appearances of the heavens, although he condemned those who did not similarly predict the time of the coming of the Messiah from the Scriptures, as is stated in Jeremiah 10 [10:21].

Moreover it is clear that Moses set no limits on human knowledge, and that through him God did not teach either physics or astronomy. For Solomon has said that "God has given the world over to human investigation" [Ecclesiastes 3:11], and Solomon himself carefully investigated all things, having studied the natural world and not just the book of Moses. The brief account given by Moses of the creation and order of the heavens, earth, and all things is an account provided by a legislator and not by a scientist. For in order to show that the God who gave him the law is the same God who created and governs the world, he began with creation, and then discussed governance, and ended with the particular type of governance through the law which was given to him.

This is proven by the testimony of all the Fathers, who also point out that Moses spoke in the language of the people and not of the philosophers, appealing to the experience of the common man and not to the philosophical mind. Therefore, since he was expert in all the sciences, both human and divine, and since he was well acquainted with all the wisdom of the Egyptians, as is said in Acts 7 [7:22] and is proven by Philo[116] and Josephus,[117] Moses satisfied the needs of both the common man and the philosophers. For those who would understand the mystical sense of the Scriptures he provides everything that must be known in both words and deeds, as is clear in the construction of the tabernacle as an image of the heavens, in the construction of the candelabra as an image of the seven planets, and in the vestment of Aaron as an image of the entire orb of the earth and of the famous deeds of our ancestors,[118] as Solomon says in Wisdom 18 [18:24], and as is proven by the rabbis and in Paul's Letter to the Hebrews [9:23].

Augustine[119] and Chrysostom[120] teach that Moses did not speak of the creation of the angels, since common people cannot understand immaterial things, and thus they might adore the angels since they were prone to idolatry. But the words creating the heavens, "Let there be light" [Genesis 1:3], can be taken by educated people to refer to the angels. Also Moses made no mention of matter to the common people, but educated people can take this to be included under the words "water" and "earth." Also Moses assigned six days to creation, but Augustine[121] along with other Fathers, takes these to be angelic not physical, days. Also St. Thomas, in Question 68 of the First Part,[122] teaches that Moses did not mention air, because he did not wish to introduce something unknown to the common people, who did not know whether air is a body, because it is invisible; but this is implied in the words, "There was darkness over the face of the abyss" [Genesis 1:2].

Thus all the Fathers who have examined the text of Moses philosophically agree unanimously that his mode of speech is directed to the capacities of the common man. Thus in his *Allegorii super Moysen* Bishop Anastasius[123] shows that Moses is to be read primarily in the allegorical sense. Even Chrysostom,[124] the greatest spokesman for the wisdom of Moses, who is rather opposed to the allegorical sense and reduces almost everything to the literal and moral senses, still admits that in this book Moses has adapted almost every word of his language to the common man.

This is especially true where Moses says that "God made two great lights" [Genesis 1:16]. He calls the moon a larger light because of its effect on us and because it appears to be larger to the senses, even though it is smaller than the earth and many stars. In Question 70, Article 1,[125] St. Thomas shows that Moses spoke here and elsewhere according to the experience of the common man, and not according to reason, which knows that the moon is smaller. St. Thomas also says the same thing about the motion of the stars and of the heavens. Since the motion of the stars is obvious to the senses but the motion of the spheres is not, Moses did not say that the latter move, which he could have said if Aristotle's opinion is true. See St. Thomas' reply to the third and fifth objections. Therefore anyone who would wish to condemn the astronomers because they think that the moon is smaller than most stars, that it is only about one-third the size of the earth, and that it does not have any light of its own, all because Moses called the moon "a greater light," would be ridiculous and wickedly ignorant, as we will soon show below in the fifth assertion.

PROOF OF THE FOURTH ASSERTION

Every human society or law which forbids its followers to study the natural world should be held in suspicion of being false. For since one truth does not contradict another, as was stated at the Lateran Council[126] under Leo X and elsewhere, and since the book of wisdom by God the creator does not contradict the book of wisdom by God the revealer, anyone who fears contradiction by the facts of nature is full of bad faith.

We all know that this is why science was forbidden by the Mohammedans. For when the Moors philosophized, many of them, having detected a fraud, wrote against the Mohammedan faith, for example, Averroës,[127] Avicenna, Alfarabi,[128] Ali Albenragel,[129] Albumasar,[130] and other philosophers and astronomers, as we have shown in our *Antimachiavellismus*.[131] This is why one king of the Moors prohibited his followers from pursuing the sciences, as Boterus reports; and the same was done by the sultans of Turkey.

Among the pagans also there was a law warning against undertaking a critical investigation of the gods. Plato warns in the *Timaeus*[132] that we should speak of the gods as legislators and of what the gods themselves would want, even though one might personally believe in only one God. And Chrysostom, in his commentary on the Letter to the Romans,[133] condemns Socrates, who knew that the gods were false, for saying when he died, "We owe a cock to Asclepius," as Plato reports in the *Phaedo*.[134] Also the Athenians persecuted Anaxagoras, Socrates, Aristotle, and other philosophers to the point of death because they inquired about the gods, which was forbidden by the law. And it so happened that they knew the truth about God, as the Apostle, Cicero, Cato according to Lucanus, and many others testify. Therefore those who maintain that the Christian religion prohibits true science and study and inquiry into physical and celestial matters either have an evil opinion of Christianity or cause others to suspect it of evil.

Indeed, if the Christian religion is completely full with every truth and is exempt from all error, then it has nothing at all to fear from inquiring thought but rather finds justification in it. St. Thomas seems to say this in Book I of his *Summa contra gentiles*,[135] and in his opusculum *Contra impugnantes Dei cultum et religionem*, when he replies to those who condemn monks for studying philosophy and the other sciences. And in Part I, Question 1, he proves this from reason and from the authority of Solomon in Proverbs 9 [9:3], "Wisdom," i.e., theology, "sends its handmaidens," i.e., the sciences, "to call us to the fortress."[136] Therefore we should not abandon the sciences, but use them to call men to the kingdom of heaven, for they are the handmaidens which truly serve the kingdom and do not oppose it.

Indeed what opposes the faith are not the sciences but the fantasies of vain philosophers, as we learn from the Lateran Council,[137] from the Second Council of Nicaea,[138] and from the articles condemned at Paris.[139] That the sciences should be prescribed and not prohibited is proven by the fact that it is said in 1 Corinthians 3 [1:24] that "Christ is the power and wisdom of God." Also in Ecclesiasticus 1 [1:1, 25] it is said, "All wisdom is from God," and "The root of wisdom is the word of God." Hence those who are Christians are wise and rational through him. For the Word of God is the highest reason, from which we are called rational by participation. Christ wishes us to be rational by being as similar to him as we can in action and in truth.

Hence those who say in general that we are wise and inquire rationally only when we accept things from other humans in a sense are not Christians and contradict Christ and diminish our likeness to him. For in effect they limit the works of divine wisdom to the understanding of only one man within a small handful, and appeal to reason in the human mind, and not in Christ, as Paul wished when he subjected to Christ all rulers and wise men and "all thought," as is said in 2 Corinthians 10 [10:5]. These are the chains around the feet and the collar around the neck mentioned in Ecclesiasticus [6:25].

They are not true Christians who would chain us to Aristotle or to Ptolemy or to some other author, as do the Averroists (including Antonio della Mirandola),[140] thinking that God has made no greater minds, or who would chain us to the words of such authors and twist the meaning of Scripture accordingly, rather than interpreting it in accordance with the nature of things, which is also the book of God, and which is far better for explaining the divine Scriptures.

The wisdom of God is exceedingly vast and cannot be confined to the genius of any one human. The more it is sought, the more it is found to contain, and we then realize that we know nothing in comparison to the numerous and marvelous things of which we are ignorant. This is the knowledge which Solomon envisioned in Ecclesiastes [8:17], and which the Apostle praises [Romans 11:33], and which Socrates found in himself.[141] Those who think that they know because they know Aristotle or because, like Galileo, they know something new about the world, the book of God, do not know the method required for knowledge. They are not truly wise unless they know that there are many more things of which they are ignorant, and that they should not stop their investigations as if they already knew everything, as is said by St. Leo and in Ecclesiasticus, chapters 42 and 43. For what we know is only a glimmer.

Therefore wisdom is to be read in the immense book of God, which is the world, and there is always more to be discovered. Hence the sacred writers refer us to that book and not to the small books of humans. And we use the teachings of the pagans insofar as they share in the rationality of Christ, the first reason. For although they do not believe in supernatural things, it does not follow that they do not participate in Christ in natural things. Thus if they say something which is good, "We should take it away from them as from an unjust owner" (as Augustine says in *De doctrina christiana*, II),[142] for although they knew the truth, they did not honor it, and therefore did not deserve to receive the supernatural faith. We recognize what is in them from Christ, but we prefer our own authors because grace perfects nature even in natural endeavors, as is taught by the Fathers and by St. Thomas in I-II.[143]

Therefore, other things being equal, Christians are more suitable than pagans for the investigation of truth, and whoever subjects himself to a pagan insults Christ. "Under every leafy tree you have prostrated yourself as a harlot," as the prophet says.[144] Jerome understands this to refer to those who prostrate themselves before secular wisdom. So in his letter to Pammachius,[145] he uses a figure from the Old Testament[146] when he says, "If you will love a foreign woman, that is, the secular science of the pagans, cutting her hair and cleaning her fingernails . . . ." The Lateran Council teaches us the same thing.

In our discussion of whether there is a need for a new philosophy,[147] we have shown that in our day when the handmaiden has taken the upper hand over her master, i.e., theology, she should be sent away, as was Hagar.[148] And since some of the sons of Israel spoke Hebrew and some spoke Ashdod, Ezra ordered them to reject foreign wives and take instead the daughters of Judah,[149] that is, the doctrine of the saints. Thus our knowledge of the world, the book of God, should be reconstructed, as I have done and as Galileo continues to do. Also St. Thomas says in Question 1[150] that the pagans are quoted in the schools of theology as testimony against themselves but not as final judges, nor as testimony against us. That they would serve as teachers, especially of theologians (which astonished Bembo),[151] is incredible. May this never happen.

Therefore those who forbid Christians to study philosophy do not understand what a Christian is. They are like the Emperor Julian,[152] who abandoned the faith and outlawed all sciences for the Christians so that theology, once separated from its handmaidens, would be unable to call men to the walls of the city of God. St. Thomas makes the same point in the opusculum already cited.[153] What should we today call those who prohibit us from studying the book of Christ, which is the world, when the name "Julianists" was used for those who wished to prohibit the monks from reading secular books? They have no support from the divine Scriptures.

Even the passages, "Do not wish to know more than is necessary," [Romans 12:3] and "Whoever thinks he is wise is a fool," [1 Corinthians 3:18] support our view and are not against us. These passages do not forbid the study of philosophy; rather they forbid the stopping of philosophy as if we already knew everything. And they forbid a wisdom which is placed above revealed doctrine by personal choice, and which is used to measure divine teachings by its own standards, as is done by pagans and heretics and those who put the lamp of the Scriptures under the bushel of Aristotle. Thus many things are said contrary to human prudence in the book of Job [32–38] and contrary to astronomy in Isaiah [47:10–14]. Nevertheless it is still true that prudence is the most divine virtue and astronomy the most useful science, as Jerome teaches in his prologue to the Bible.

Therefore human prudence is forbidden when in Machiavellian[154] style it exalts itself above divine prudence and when it thinks that it can attain what is above nature by itself and not by a request from God. Likewise astronomy is forbidden when it places itself above the prophets in Babylon and presumes to predict future contingents with certainty. But it is not forbidden when it is subjected to the prophets and deals with the future moderately and hypothetically. And the same is true of the other sciences.

Appendix: It is to the glory of the Christian religion that it permits us to make the effort to discover new sciences and renew the old ones, so that we do not need "to cut the hair and the fingernails of foreigners." And lest Machiavelli and Julian continue to scoff at us, the Christian religion also makes us observers of Christ, the wisdom of God. And yet we still beg from unbelievers the sciences which we have condemned, thus all but making them superior to us. We have discussed this argument above, when we appealed to Augustine, and more fully in our *Antimachiavellismus*,[155] where we have added that the approval of the sciences by Christianity is one of the strong bonds which holds me fast in the Church of God. And I believe that this is also true of others. Why should we now break that bond?

PROOF OF THE FIFTH ASSERTION

We have proven that the freedom to engage in philosophical inquiry is more vigorous in Christianity than in other religions. As a result, if there is anyone who chooses on his own to prescribe rules and limits for philosophers as though they were decreed in the Scriptures and who teaches that one should not think differently than he does, and who subjects and confines the Scriptures to one unique meaning either of his own or of some other philosopher, then that person is not only irrational and dangerous but is also impious. For he exposes the Sacred Scriptures to the mockery of the philosophers and to the ridicule of pagans and heretics and thereby prevents them from listening to the faith. He does not summon infidels to the refuge of the faith, but he repels them. And he also offends the Holy Spirit, whose most pregnant and most fruitful words are thereby rendered quite sterile (as is attested by Augustine in his *De doctrina christiana*,[156] by Chrysostom in his commentary on the Psalms, by Ambrose and Origen in all of their works, and by Gregory in his *Moralia*, Book 15).[157] But the most fruitful meaning of Scripture is located not just in its mystical sense but also in its literal sense, as is taught by Augustine in *De Trinitate*, I,[158] by St. Thomas in I, 1, 10, and by Cardinal Cajetan[159] in his comment on the latter passage. For all meanings and interpretations of Scripture which do not directly or indirectly contradict other passages of Scripture are permitted, as is said in Question 32, Article 4.[160]

Furthermore in Opusculum 10, Question 18, St. Thomas gives the same reason for such multiple interpretations which Augustine had given earlier in Book I of his *De Genesi ad litteram*.[161] Thomas says, "The words of Sacred Scripture are interpreted in many ways so that they may take refuge from the mockery of those who are inflated with secular learning."[162] In his book *De Trinitate* Thomas teaches that this happens so that Scripture may be protected in various ways from the quibbles of heretics.[163] Also, in the introduction to that same Opusculum, Thomas writes:

> Here at the beginning I wish to declare that many of these articles do not pertain to the doctrine of the faith but rather are philosophical teachings. Now it is very harmful to affirm or deny things which are not related to the doctrine of the faith as if they actually do pertain to sacred doctrine. For Augustine says in his *Confessions*, Book V, "When I hear a Christian who is ignorant of these matters (i.e., what the philosophers have said about the heavens and the stars and the motions of the sun and the moon), and who makes a mistake, then in my patience I see a man with an opinion, but I do not see anything evil in him, as long as he does not believe anything unworthy of you, the Lord and creator of all things, even though by chance he is ignorant of the state and condition of creatures. But he is evil if he thinks that this opinion pertains to the doctrine of the faith and if he dares to affirm persistently that of which he is ignorant."

St. Thomas continues,

> That this is definitely harmful is made clear by Augustine in Book I of his *De Genesi ad litteram* where he says, "Something which is especially ugly and dangerous and

to be avoided at all costs is the situation in which an infidel is hardly able to restrain himself from laughing when he hears some Christian foolishly speaking of something or other as pertaining to the Christian faith while the infidel knows that what was said is completely false. The troublesome thing is not that the man has been caught in an error, but rather that those who are outside the Church would think that our authors believe such things, and would thus condemn us as ignorant. And hence great harm has been done to those whose salvation is a concern for us. The safer course seems to me to be the following. Whatever the philosophers generally take to be true, and which is not contrary to our faith, should not be affirmed to be a doctrine of the faith, even though it is sometimes so presented under the name of philosophy; and it also should not be denied as contrary to the faith, lest this provide an occasion for the wise men of this world to scorn the teachings of the faith.[164]

So says St. Thomas in agreement with St. Augustine. From this it is clear that modern authors who defend Aristotelianism as though it were part of the faith are both ignorant and in disagreement with the Fathers. They do this because Thomas has written commentaries on Aristotle, even though here Thomas teaches exactly the contrary, as we will see more fully in our replies[165] to the arguments. For example, Ulisse Albergotti maintains that the moon shines by its own light, because the Scripture says, "The moon will not give its own light" [Ezekiel 32:7], giving a forced reading to the words "its own," which however can be interpreted in many ways.[166]

But it is surprising that both Augustine himself and the Fathers made a similar mistake, namely, an error in the particular premise, but not in the universal premise which they were teaching, in the following syllogism.[167] First, Lactantius Firmianus in Book 3, chapter 25,[168] and later Augustine in *De civitate Dei*, XVI,[169] firmly conclude that the antipodes do not exist, because those humans would not be descendants of Adam, which is contrary to the Scriptures which say that the whole human race has descended from one man. They also gave physical arguments. In 500 A.D. Procopius of Gaza[170] proved that the antipodes do not exist by drawing up a list of comments on Sacred Scripture taken from the writings of all the Fathers. Using this list and the authority of the Sacred Scripture, St. Ephrem located the terrestrial paradise in the whole of the other hemisphere discovered by Columbus. Indeed some of the Fathers who have held that the antipodes do exist have even been taken to be heretics. Nevertheless the navigators have shown that the denial of the antipodes is false. Therefore if the existence of the antipodes is truly contrary to the Scripture of God, as some have said, and if either the terrestrial paradise or purgatory and hell are located down below, as Dante,[171] Isidore,[172] and others have thought, then it follows that the truth recently discovered by Columbus would contradict, or at least be in disagreement with, the divine Scriptures.

Furthermore Procopius and others think that the earth was established on, and floats on, water, as the philosopher Xenophanes[173] once said. And they prove this from the Scriptures, where David says in Psalm 135 [135:6], "He set the earth upon the waters," and in Psalm 23 [23:2], "He founded it upon the seas." Nevertheless it is now apparent that the earth is suspended in the middle of the world,[174] sustaining both itself and the waters, and that it is not held up from below by the waters, as they believed.

In nature there is no "below." There are rather centers, determined by the conservation of each system, as its parts tend toward the center so that the unity and preservation of the whole will be maintained. Thus the parts of the sun tend towards the center of the sun, and the parts of the moon tend towards the center of the moon. St. Ambrose[175] worried that the motion of the heavens was composed of a rising and a falling, so that it would come to a stop in time, as was thought by Chrysostom and other Fathers. But these arguments are of little importance in astronomy.

Consider how dangerous it would be to affirm that such things are part of the faith. Bishop Philastrius[176] pronounced things to be part of the faith which actually are not—for example, that the world is exactly as old as he himself said it was and that when God instilled the breath of life into Adam, he gave him the Holy Spirit rather than a soul. These two claims are ridiculous for both Catholics and heretics. Bede[177] was more cautious when he said that dropsy is an illness caused by a failure of the bladder. And St. Thomas was more cautious when he said, persuaded by the authority of Aristotle, that humans cannot live in the tropics,[178] even though Albert and Avicenna thought otherwise. For neither of them said that these claims were part of the faith, even though St. Thomas could have appealed to the "flaming sword."

Today geography and medicine have shown that they were wrong, but this is no threat to the faith. Much worse is the error of those who claim that the torrid zone is the "flaming sword" of the angel who guards the road to paradise,[179] for by now travelers and navigators in the tropics have found that there is no such barrier. What would pagans and Mohammedans say when they hear that we find such things in the Scriptures? We could reply to the Mohammedans that they believe that under this earth there are seven other earths, and an ox, and a fish whose head faces east and whose tail faces west, holding up all these worlds. But it is a small consolation to display the errors of others when and where we ourselves are in error.

For these reasons then, if Galileo wins out, our theologians of the Roman faith will be the cause of a great deal of ridicule among the heretics, for his theory and the use of the telescope have by now been enthusiastically accepted by everyone in Germany, France, England, Poland, Denmark, Sweden, etc. But if Galileo's view is false, there is no difficulty at all for the teachings of theology. For in the Church Militant not every falsehood is contrary to the faith, even though it perhaps is contrary to the faith in the Church Triumphant. Otherwise errors made by the saints in physical matters would prove them to be heretics. Furthermore, if falsity is found, his view will not last.

Therefore I think that this philosophical theory should not be condemned. One reason for this is that it will be embraced even more enthusiastically by the heretics and they will laugh at us. For we know how greatly those who live north of the Alps complained about some of the decrees adopted at the Council of Trent. What will they do when they hear that we have attacked the physicists and the astronomers? Will they not immediately proclaim that we have done violence to both nature and the Scriptures? Cardinal Bellarmine[180] is well aware of this. Another reason is that both Augustine and Thomas think, as was proven, that it is permissible, and ought to be permissible, to say that the heavens are composed of the fifth element and to name the days of the week after the dominant planets, as St. Thomas says in Opusculum 10, Article 39, in agreement with what he established in his introduction.[181]

PROOF OF THE SIXTH ASSERTION

The sixth assertion needs no further proof. For it is evident that falsity is not contrary to Catholic teaching unless it clearly disagrees, either directly or indirectly, with Sacred Scripture or with a decree of the Church. And it is also evident that in such cases we should suspend assent and not pass rash judgments within the Church, as St. Thomas and Augustine have said, as quoted in the fourth assertion. From what has been said it is clear that the teachers of theology have embraced many errors from the pagan philosophers; for example, Xenophanes' view that the earth floats on water; and that the antipodes do not exist; and that at night the sun moves across the northern part of the earth, but is not seen because of the mountains, as Aristotle relates in Book II of his *Meteorologica*;[182] and that the torrid zone is uninhabitable; and that the terrestrial paradise is located in the islands of the blest, or in the Orient in China, or on the moon; and other such things. However after the error of these views was detected, they did not become heresies.

But falsity cannot be found in Galileo, because he does not proceed from an opinion but from sensory observations of the book of the world. He does not speak in terms of the faith, which would expose both himself and the Scriptures to ridicule if an error were found. I will discuss this further in the replies[183] to the arguments, where I will also show, without any embarrassment to the faith, how much more dangerous are Aristotle's arguments.

THIRD HYPOTHESIS

Whoever would wish to be a judge in this case must understand that our previous remarks are fundamental. And since the present dispute concerns the physical knowledge contained in the Sacred Scriptures, whoever wishes to be a judge must, as said earlier, thoroughly understand the methods of explaining all the literal and mystical senses of Sacred Scripture according to the commentaries of the holy Fathers, and must also understand the book of nature as found in all the sciences and especially the observations made by physicists and astronomers.

Scripture, which is the book of God, does not contradict the other book of God, i.e., nature. But reading the book of nature requires a very observant person who is so well versed in all the sciences that he can examine the apparent disagreements and the hidden agreements between the two books. The books should be interpreted in the light of the teachings of all the philosophers and not just Aristotle or any other single philosopher. And as we read each book of God, we should explain its proper meaning according to the spirit of the Fathers and the most fruitful understanding of the holy Church, and we should be free of all jealousy and passion, which clouds and distorts our judgment. We do not want to be included among those judges mentioned by Horace who lost the respect of their contemporaries, "Either because they thought that only what pleases them is good, or because they thought it shameful to agree with the young and yet to admit that the old should reject what the young have to teach."[184]

St. Jerome in his *Epistola ad Magnum*, after declaring that the sacred writers were acquainted with the teachings of all the philosophers, adds, "I pray that you persuade them" (i.e., those who disagree with the above statement), "that the toothless should not envy the teeth of those who eat, and the mole should not condemn the eyes of the goat."[185] For out of envy they turn themselves into trouble-makers for more profound modern thinkers, because they themselves are ignorant of such things, or despair of learning them, or are embarrassed to become students now that they are called teachers.

CONCLUSION OF CHAPTER 3

We have proven that neither zeal for God without knowledge, as Bernard said, nor knowledge without a zeal for God, is enough to judge this case. We have also shown what must be known, and how one must be zealous for God, not for man, keeping before our eyes the passage in Numbers 11 [11:29] where Joshua, who in his zeal for Moses did not want there to be other prophets in the encampment, heard Moses say, "May it be that all the people are prophets, and that the Lord might give his spirit to them." For a much better reason St. Thomas could now say the same thing about himself.[186] Hence it is rather embarrassing that, in our crude zeal for Aristotle rather than for Moses or St. Thomas, we prohibit our fellow Christians from philosophical inquiry and give our preference to pagans.

CHAPTER 4

Replies to the Arguments against Galileo Stated in Chapter 1

REPLY TO THE FIRST ARGUMENT

We have already replied to the First argument against Galileo in an earlier treatise[187] in which we discussed whether it is permissible to establish a new philosophy and to strip the Peripatetics of their authority. Here we will only briefly point out that it is heretical to say that theology is based on Aristotelianism or that it essentially needs the teachings of the philosophers to prove itself. Rather theology uses Aristotle only for our sake, not as a judge in theology, nor as evidence against us, but as evidence against the pagans and other sophists, and only when philosophy provides not its opinions but its testimony about what it sees in this world. This was proven in our Second Hypothesis by means of an appeal to St. Thomas' *Summa theologica*, Part I, Question 1, his *Contra gentiles*, I, and his Opusculum 10.[188]

When St. Thomas was seen to violate his own rules about too much allegiance to Aristotle in his writings on theology, he was reprimanded for this in the Articles of

Paris. But he can be excused as blameless on this, as we have shown in our earlier treatise just mentioned above. Also, whoever condemns Galileo because he disagrees with Aristotle should first condemn Augustine, Ambrose, Basil, Eusebius,[189] Origen,[190] Chrysostom, Justin, and other saints and doctors of the Church, who condemned almost all of his teachings in physics and metaphysics, preferring instead Plato and the Stoics, as is clear when you read them. St. Justin, who is called philosopher and martyr, wrote a book entitled *Contra Aristotle*.[191]

Those who think that the downfall of Aristotle will have only a small effect on theology do not know what they are talking about, and they are committing an impious error (as was proven in the fifth assertion of the Second Hypothesis). We have proven the opposite.[192] Unless Aristotle's authority is overthrown, his heresies will continue to plague us; for example:

1. That motion is eternal, for otherwise there would be no God. Aristotle firmly maintains this in *Physics*, VIII, and in *Metaphysics*, XII,[193] which St. Thomas attests in Lecture 10[194] of his commentary on the latter work and which he rejects along with Justin and other Fathers.

2. That the soul is mortal or that there is only one immortal soul for the whole human race.

3. That God has no concern for creatures.

4. That God moves heavenly bodies in the opposite direction in which the angels move them.

5. That there is no reward or punishment after death.

6. That hell is a fiction.

7. That God acts out of necessity.

8. That chance violates the order of providence, and many other things contrary to faith, as is attested by St. Thomas, Averroes, Alexander,[195] and other Greek and Arabic writers.

Because of this St. Vincent[196] and Don Serafino da Fermo[197] in his commentary on the Apocalypse called Aristotle "the cup of the anger of God" which is poured out by the third angel over the waters of wisdom [Revelation 15:7; 16:4]. And in his book *Contra Celsum*[198] Origen says that Aristotle is more evil and more impious than Epicurus. Notice also how often and in what ways Aristotle is criticized by Augustine, Ambrose, and Justin, who understand him on his own terms.

I am astonished that there are fools who think that theology is based on Aristotle. Some of my fellow monks even attribute this view to St. Thomas and praise him for it, just as the theologians of Paris in his own day thought the same thing, but criticized him for it. But St. Thomas professed and maintained just the opposite. In our treatise mentioned above[199] we have explained why he wrote commentaries on Aristotle, and how he used Aristotle for the good of the faith by transforming a poison into an antidote.

But Galileo stands firmly on the foundations of the faith. And he speaks carefully about natural things, according to the evidence of observations, and not as someone merely offering opinions such as Aristotle spun out of his own mind. For this reason Galileo should be praised. For the overthrow of the teachings of infidels and the errors of pagans is not the destruction of theology, but the strengthening of Christianity. This is one of the things that a judge ought to know, as we have said earlier. We have shown elsewhere,[200] with the help of Nicephorus and other Church historians, how heresies blossom forth out of Aristotelianism, how and why the Averroistic version of Aristotle is the workship of Machiavellianism, and why a philosophy which is derived from God's book of the world and which serves as the handmaid and witness for theology, is not derived from the views of Aristotle or any one else.

REPLY TO THE SECOND ARGUMENT

I reply to the second argument by denying that the teachings of Galileo are opposed to all the scholastics and to

the Fathers. For even though his teachings do not agree literally with some of them, still they agree in intention. For the
Fathers wished to have the truth presented to them; however they did not speak in philosophical matters as direct
witnesses but as men who quoted the opinions of others.
Therefore in such matters preference should be given to direct witnesses rather than to the Fathers, just as today
Christopher Columbus is preferred over Lactantius, Procopius, Ephrem, and other holy doctors, and as Magellan is
preferred over St. Thomas, Anthony, and others.

Moreover in support of this I will prove first, that
some theologians have embraced philosophical teachings
which are more opposed to Scripture and to the holy doctors than are the teachings of Galileo; second, that many of
the Fathers and scholastics agree with Galileo; and third,
that Scripture agrees more with Galileo than with his
adversaries.

Proof of the first point. That the heavens, and especially the stars, are not composed of a fifth type[201] of
matter, but of the four elements or perhaps only of fire,
was once taught by all philosophers and by Saints Augustine, Ambrose, Basil, Justin, Cyril, Chrysostom, Theodoret,[202]
Bernard in his sermon "Mulier amicta sole,"[203] and by the
Master of the Sentences.[204] In Book IV of his *Hexameron*[205] Ambrose proves this from Scripture, where it is
said, "The heavens will perish and all will wear out like a
garment." [Psalm 101:26–27; Hebrews 1:10–11] Philoponus[206] says the same thing when he explains the books
of Aristotle's *De caelo* against Aristotle and in favor of
the Christians.

Nevertheless, without being condemned, as they say,
by the Scriptures, many scholastics say that the heavens are
composed of a fifth type of matter. But Ambrose has attacked this in innumerable places as an imaginary and dia-
bolical invention, as have Justin and Basil also. And when
St. Thomas discusses Aristotle in Part I[207] where he explains the text of Moses concerning the work of the six

days of creation, he first outlines both views, i.e., the opinion of the philosophers and the Fathers and the opinion of Aristotle. But then in Questions 65, 66, 67, 70, and 71, he consistently teaches that the former agrees better with the text of Scripture while Aristotle's view is in disagreement, which only a fool would not notice.

Moreover the Scripture of God testifies that the sun in itself is most hot and most luminous. For Genesis 1 [1:16] speaks of "the greater light;" and the sun's heat is mentioned in Psalm 18 [18:7], in Wisdom 2 [2:3], in Ecclesiasticus 43 [43:2–4], and many other places; and Wisdom 17 [17:5] speaks of the illumination of fire and of "the brilliant flame of the stars." That this is the way things are, and that to think otherwise is heresy, is taught by Ambrose in Book IV of his *Hexameron*, and Basil agrees, as also do Augustine, Chrysostom, Justin, Bernard, Origen, Philoponus, and all the Fathers I have ever read. In the Church's Ambrosian hymn we sing, "Iam sol rededit igneus."[208]

Nevertheless other Scholastics think that the sun in itself is not hot, without being marked as heretics, and the Church does not prohibit this opinion. Even Aristotle himself, the author of this theory, says that there is no light in the sun, as is clear in *De caelo*, II, 7,[209] where he teaches that light and heat are produced by the friction of the air, which both Simplicius and Alexander agree is Aristotle's view. In his *De substantia orbium coelestium* Averroës states that Aristotle denied that there is any light or heat in the sun. More recent authors, who do not accept Aristotle's teaching, have restored light to the sun. But indeed if light should be attributed to the sun so that Scripture would not be contradicted, then so also should heat be attributed to the sun. However Aristotle denied that there is light in the sun, for otherwise it would be composed of fire.

Nevertheless many modern writers disagree with Aristotle and the literal meaning of Scripture, and they are not condemned. Should Galileo then, who proves his teachings from sense evidence, be prohibited from observing the book of God? I will omit discussion of other views which long ago were held to be part of the faith but which are now known by common experience to be false; for example, that the antipodes do not exist; that the land near the equator is uninhabitable; and that heaven or hell is located in the other hemisphere or in the islands of the blest. I will pass over the fact that Procopius, Eusebius, and others maintained from Scripture that the earth floats on water, while others who held the contrary were not condemned, and now experience has shown that the former were wrong. These facts are in Galileo's favor.

Proof of the second point. To begin with, the question of whether or not the earth is in the center of the world is in no way a dogma of the faith, as St. Thomas was quoted in our fourth assertion, and as was stated by the Fathers and later by the Scholastics. For in the first place Lactantius in Book III, chapter 23, Procopius, Diodorus Bishop of Tarsus,[210] Eusebius Bishop of Emissa, Justin in his *Questiones ad orthodoxos,* and others have taught that the earth is not in the center of the world and that the heavens are not spherical. Chrysostom holds the same view in his homilies 6 and 9 on Genesis; and in his homily 31 on the Letter to the Romans he says that humans do not know where hell is, which Augustine also taught in *De civitate Dei,* XXII, 16, and Peter Lombard in *Sententiae,* IV, 44, and St. Thomas in Opusculum 11, Article 25.[211]

The fact that hell is located in the center or in some other part of the earth can be deduced from the word "inferno" [hell] and from what the Apostle says in Ephesians 4 [4:9], "Christ descended to the lower regions of the earth." Hence hell is there, unless we assume that there are other earths. And David, speaking of Christ descending to the lower regions, says, "You will not abandon my soul in hell" [Psalm 15:10], as the Apostle Peter explains in Acts 2 [2:27]. Therefore we do not know whether the earth is in the center of the world.

If someone were to hold that the infernal darkness, which Christ called "exterior,"[212] is located outside this world, as Origen conjectured in his commentary on Matthew[213] and as Chrysostom thought in his commentary on the Letter to the Romans, then there are other worlds outside of our world. The censors condemned this idea in Galileo because they had not carefully examined the Scriptures and the books of the holy Fathers. But what is beyond controversy is that in his homily 7 on the First Letter to the Thessalonians,[214] Chrysostom says that we can know that the earth is cold, dry, and dark, but nothing more, and especially its location and situation in the world, etc. Therefore Scripture does not teach us what is at the center rather than at the circumference.

Furthermore Chrysostom teaches that we do not know whether the earth is in motion or at rest, for he limits what we can know about it to the three conditions mentioned above, i.e., coldness, dryness, and darkness. Agreeing with Chrysostom are Theophylactus,[215] Lactantius, Augustine, Procopius, Diodorus, Eusebius, and others; and Justin maintains that the earth is not in the center. Therefore I do not understand why our theologians today, without any prior mathematical proofs or experience or revelation, think that they know for certain that the earth lies motionless in the center of the world, and that to think otherwise contradicts the Fathers and the Scholastics, whom they have not read.

If indeed it is true that hell is located in the center of our earth where the damned are tormented by fire, as Gregory and others seem to think, then it must be that the earth moves. For as Aristotle reports,[216] Pythagoras, who located the place of punishment in the center of the earth and who thought that fire is a cause of motion, concluded that the earth moves and is animated, as was believed also by Ovid in Book XV of his *Metamorphoses*,[217] by Origen in his commentary on Ezekiel,[218] by Alexander of Aphrodisias, and by Plato. But in Opusculum 11, Article 24,[219]

St. Thomas says that this is contrary to nature and would require a miracle, and so he thought that hell is located in some other unknown place. Therefore if hell is in the center of the earth, then for the reason given by St. Thomas, the earth is hot in its interior and it moves, as Gregory[220] and others have thought. Hence Galileo's view is not inconsistent with St. Gregory but with Aristotelianism.

Procopius, Diodorus, Eusebius, and Justin teach that the starry heavens are immobile. In his homily 12 to the people of Antioch, in homilies 14 and 27 on the Epistle to the Hebrews, and in the work quoted above, Chrysostom proves from Scripture and from reason that the heavens are immobile. For the Apostle, speaking of the heavens as the tabernacle of Christ the priest, says in Hebrews 8 [8:2], "It is fixed by God, not man," where "fixed" means that the heavens are immobile, and he says the same thing in chapter 12 [12:26]. In his *De Genesi ad litteram*, II, 10, Augustine says that the mathematicians of his day have shown with certain proofs that the heavens are immobile, and that in such matters we should neither refute nor strongly embrace the philosophers, lest we expose theology and ourselves to ridicule.

Moreover the above mentioned Fathers think that the heavens are neither spherical nor in motion, for the contrary is denied by the prophets, by Moses, and by Psalm 103 [103:2], which, in the version used by Chrysostom, says, "He established the heavens like a dome, and spread them out like a tent." Justin records that this was at one time a topic of controversy between Christians and the pagans. Copernicus explains this etymologically by saying that the heavens "engrave"[221] all things. We agree with Basil, who pointed out that the heavens are spread out by heat. Moreover Bede and Strabone[222] and the Fathers mentioned above say that the starry heaven is what Moses called the "firmament," and from this word they argued that it is firm and stable. Also Paul says, "God fixed it" [Hebrews 8:2], and David says, "The heavens are fixed by the word of God" [Psalm 32:6]. If the moderns contradict this, they are not thereby heretics.

Hence it is astonishing that anyone would think that the teachings of Galileo are contrary to all the Fathers when the Fathers themselves accept that contrary view. That this was their common opinion is attested to by Sixtus of Siena[223] in his *Bibliotheca sancta*. Finally in his *Sententiae*, II, 14, Peter Lombard, the teacher of all the Scholastics and one thoroughly acquainted with the doctrines of the Fathers, says, "The Holy Spirit did not wish to tell us the structure of the heavens." Next he asks whether the heavens are fixed or in motion. And he says that both views are consistent with the Scripture; the former because the heavens are called the "firmament;" and the latter because it is the stars which are seen to move in the heavens, not the heavens themselves, and thus it could be that the heavens are at rest and the stars move, granting that the latter are not like knots in a board. And thereby one can satisfy both vision, which sees the stars move, and the biblical text, which asserts immobility. Therefore he leans towards Galileo's view. Hence it is not the case, as the adversaries claim, that the Fathers and the Scholastics have maintained that the earth is at rest and the heavens move.

Proof of the third point. In his commentary on *De caelo*, II, from Lecture 20 up to the end, St. Thomas discusses Aristotle's opinion about the motion of the earth and the stability of the heavens, but he nowhere says there that this is contrary to the Scriptures, which he is regularly accustomed to do in regard to other teachings of Aristotle and the philosophers. The reason for this is that in that work his purpose is to explain the text of Aristotle.

But in Opusculum 10, Article 16,[224] where he has the opportunity to say whether this is contrary to Scripture (for the question at hand there was whether the earth has a rotational motion or can be moved by an angel), he says that this is contrary only to Aristotle and not to Scripture. For provided that the order of the universe established by

God does not change, then individual opinions about the situation and motion of its elements are not contrary to Scripture, as St. Thomas wisely maintains. Moreover all theologians who say either that the heavens are at rest or that it is not contrary to the faith to say so, must also say as a necessary consequent that the earth moves or that it is not contrary to the faith to say so. Included among these theologians are the Master of the Sentences, Chrysostom, Lactantius, Procopius, and Augustine. Thus Sixtus of Siena rightly maintains that the stability of the heavens is not contrary to Scripture, as the uneducated think.

Saints Bede[225] and Strabone assert that the firmament is the same thing as the starry heavens. Those who think that the firmament is something else (being unable to explain how the stars are seen to move when they are located in the firmament, which the Master explained by saying that they are not fixed there like knots in a board) are forced to say that the firmament contains no stars and is outside of the heavens which we see, even though Moses located the stars in the firmament. Hence Bede and Strabone, who are commended by St. Thomas, have captured the meaning of the Scripture better by saying that the firmament is the same as the starry heavens.

When the Fathers, in order to save the apparent motion of the stars, say that the stars move but not the firmament, this can be true only of the planets, as St. Thomas and Sixtus explain. Otherwise this would be absurd, as St. Thomas realizes. For there are innumerable stars in the firmament, especially in the Milky Way, and in their constellations they always preserve the same relation, order, and motion in relation to each other, even though they vary in latitude and location in relation to the equator and to the zodiac. But it would not be possible to maintain a perpetual order in such a multitude, for since some of the stars are closer to the earth and some are further away, this causes a parallax which changes the order of their position, at least for those who observe them. Furthermore the stars, like the planets, cannot all move with the same motion, because some are large, while others are small, and they thus have unequal powers. Hence they undergo various motions, just as the planets have various motions because of their different sizes and powers, whether they are moved by themselves or by the sun. Simplicius uses the same argument to prove that the heavens are not composed of fire, for otherwise the stars would move like fish in the sea with much variation and inequality, and would not always be in the same place. But our argument is better than Simplicius'.

Therefore if according to the Fathers the firmament is motionless, then so also are the stars within it. And thus since Chrysostom, the Master of the Sentences,[226] and the other Fathers mentioned above, maintain that a motionless firmament conforms to the Catholic faith, then much more should they say the same thing about the stars. Hence it follows that the earth moves in a circle, like a ship, and the stars appear to move in a circle, like an island or a tower on the shore.

This then is the cause of the appearances. And this is consistent with the Scriptures which speak of the stability of the firmament, in which God placed the stars, without any distortion or absurdity. St. Thomas discovered this but, as was his custom, he concealed it out of respect for the Fathers, as he himself indicates in Opusculum 1.[227] Therefore the Fathers and the teachers of the Scholastics, Thomas and Peter Lombard, are more in support of Galileo than against him, and the Scriptures favor them more than Galileo's censors.

REPLY TO THE THIRD ARGUMENT

To the third argument I respond that when the Psalm [92:1] says, "He established the orb of the earth," this refers to its location and order within the whole, which will remain perennially fixed and unchanged. And likewise when the other Psalm [103:5] says, "He established the earth above its foundations; it will not move from age to age," this refers to all time up to, but not including, the end of the world when "the heavens and earth will tremble," as the Church says in its hymn[228] with the prophet.

The adversaries cannot reject this interpretation. For those who claim that the firmament moves reply to the argument by saying that the heavens are called "firmament" because their motion maintains the same constant order. Moreover in Job [37:18] we read that the heavens are "most solid, like a bronze spindle;" and yet Basil[229] teaches on the authority of Isaiah that they are composed of the most subtle fire, and he explains their solidity differently.

Furthermore, in favor of Galileo and Chrysostom is the passage of Proverbs 8 [8:27–28] which says, "When he prepared the heavens, I was there, when he surrounded the abyss with a fixed law and circle, when he established the aether above, and when he balanced the fountains of the waters." Here we see that God established the aethereal heavens and balanced the fountains, as we will see further below. And David says, "By the word of God the heavens were made" [Psalm 32:6]. Therefore the Scriptures speak no more about the stability of the earth than about the stability of the heavens. As a result neither those who say the heavens move nor those who say the earth moves conflict with the divine Scripture, which allows for both meanings.

Furthermore, in Psalm 135 [135:6] it is written, "He established the earth above the waters." But Galileo's enemies deny that "above the waters" is to be taken literally; and with equal justification Galileo denies that "established" is to be taken literally, but rather this is said only according to the appearances. But when it is said, "The earth remains forever" [Ecclesiastes 1:4], without doubt this refers to its condition in regard to destruction. For as Solomon says, some things die and some are born, but the earth remains and will never perish in total ruin but will persevere always in its proper state.

If hell is in the center of the earth, which is the common opinion believed by many, then (if St. Thomas' argument is valid) it is necessary to say as a matter of faith that the earth, which is hot in its center, is located outside the center of the world and is in motion, for such is the nature of fire and animated things. In Opusculum 11, Article 24,[230] St. Thomas maintains that hell is not in the center of the earth, because he thinks that the earth is cold, that all heavy bodies in the world tend towards its center, and that the world is not deprived of a purpose. Further, there is no question of a miracle in this case, because we are dealing with the world's constitution ("Tophet[231] was prepared beforehand," as is said in Isaiah 30 [30:33]), and thus with the natural, not the miraculous, as Augustine says.[232] From all this therefore St. Thomas ought to say either that the earth is hot and also has motion on its surface, or that hell is not in the center of the earth.

REPLY TO THE FOURTH ARGUMENT

To the fourth argument I say that it is clear from what has been said and from the text of Solomon that the word "stands" here exempts the earth from the change of corruption or destruction, but not from local motion. For the full text [Ecclesiastes 1:4] reads, "A generation passes, and a generation comes, but the earth stands forever," that is, it will not be destroyed. In contrast it is said in Job [14:2] that corruptible humans "never remain in the same state."

What is said next in the text of Solomon, about the rising and setting of the sun and its rotation through the north, has received many interpretations without any damage to the Scriptures. Augustine,[233] Lactantius,[234] and others say that the sun does not rotate under the earth but along its northern side, where we do not see it because of the great mountains. This view was held by Xenophanes and the ancient philosophers who denied the existence of the antipodes (as Aristotle relates in *Meteorologica*, II),[235] and Augustine agrees with them because he likewise did not accept the antipodes. In Opusculum 10, Article 18,[236] St. Thomas interprets this text in agreement with Ptolemy to the effect that an angelic spirit moves the sun. He says the same thing in Opusculum 11, Article 6,[237] where he adds that the more scriptural interpretations we have of this text and others, the better is Scripture protected from the ridicule of secular philosophers.

As a result I would prefer to interpret this passage differently to avoid the ridicule of the Germans, who now take it to be certain that the earth moves and the sun is at rest at the center; namely, Copernicus, Reinhold, Stadius, Maestlin, Rothman, Gilbert, Kepler; and numerous Englishmen and Frenchmen; and the Italians Francesco Maria Ferrarese, Giovanni Antonio Magini, Cardinal Cusa, Colantonio Stigliola,[238] and others, as I have said in chapter 3, Second Hypothesis; and among the ancients Pythagoras and all his followers, Heraclitus,[239] Aristarchus,[240] Philolaus,[241] etc., whose opinion St. Thomas did not condemn as heretical, as I have shown in my reply to the second argument. And even if he had condemned it, it would not thereby be heretical.

St. Chrysostom says that it is heresy against Scripture and the Church to say that there are many heavens and spheres. Philastrius says[242] that those who do not agree with him about the age of the world are heretics. Ambrose thinks that it is heresy to say that the sun is not intrinsically hot. A modern writer[243] says that it is heresy to say that the moon shines by means of borrowed light. And Procopius says it is heresy to deny that the earth floats on water. But those Scholastics who have an opposite opinion on all these matters are not thereby heretics, for the Church has not defined these teachings and the Scriptures are open to many interpretations.

Therefore it is permissible for Galileo to interpret this passage differently; namely, that the sun only appears to move relative to our senses. In Part I, Question 70, Article 1, Reply to Objection 3,[244] St. Thomas likewise says that Moses spoke of these matters according to sensible appearances, using the popular and not the philosophical meaning. And in the same place, in Reply to Objection 5, he responds in agreement with Chrysostom[245] that Moses calls the moon a "great light" because of its effect relative to us and to our senses, for many stars are larger than the moon. Indeed someone living on Jupiter would say, "God created five great lights: the sun which is the greater light, and four smaller lights, i.e., the Medicean stars." For these moons of Jupiter would appear as large to an inhabitant of Jupiter as our moon appears to us who live on the earth. The astronomers say that four moons revolve around Jupiter, and two around Saturn,[246] just as Venus and Mercury rotate around the sun. The whole of Scripture accommodates its words grammatically and in their meaning to the senses of the people (as we have proven in our Second Hypotheses from the testimony of Chrysostom, Augustine, Thomas, Origen, Bede, and all the Fathers).

God handed over the world, his first Scripture, to the examination of everyone; and he handed over his second Scripture to the learned to examine its various meanings, but within the limits set by the Church. Thus Christ, the incarnate wisdom of God (as Origen says), shows himself as a man to children and the uneducated, as a prophet to the learned, and as God to spiritual people. And the world, which is the materially created wisdom, shows itself in many ways and under many capacities. And so Scripture is the written wisdom.

REPLY TO THE FIFTH AND SIXTH ARGUMENTS

I respond to the fifth and sixth arguments by denying that the two miracles of Joshua and Hezekiah would be nullified if the sun is at rest in the center of the world. The text says that the sun stopped and turned backwards relative to our senses; but it would be truly miraculous if the earth actually stopped and moved backwards. For the stopping of the sun is not a greater miracle than the stopping of the earth. In the words of the Church's hymn, "Now the sun sets like a flame,"[247] you would interpret the word "flame," without any fear of heresy, to have an equivocal meaning which is relative to us. Likewise in the words of the Scripture, I would interpret the phrase "stopped and turned backwards" to have an equivocal meaning which is relative to us, and is not really what happened; just as Virgil says, "We set sail, and the city and the shore receded,"[248] even though it is we and not the city which receded. Indeed my interpretation agrees better with the text.

The hymn mentioned above was composed by St. Ambrose, as is clear from the last volume of his writings. But Ambrose says in his *Hexameron*[249] that the sun is formally and intrinsically composed of fire, that anyone who thinks otherwise is a heretic or a fool, and that the Aristotelians are to be strongly criticised on this point, as was done by the Fathers quoted in our reply to the second argument. Like the creed of Athanasius,[250] the Church has adopted this hymn as its own, a hymn which is full of meaning and which is open to many interpretations. Nevertheless the Church does not say that those who deny that the sun is intrinsically hot are heretics, for it does not condemn those who say this.

One would indeed contradict Scripture if one were to say that what happened in the cases of Joshua and Hezekiah were only sensory illusions and not true miracles. We agree that these were true miracles; but the appearances are exactly the same if either the observer or the object seen is moved, as optics teaches us. Moreover these are miracles for us, not for God, for whom nothing is miraculous. And miracles occur not for God's benefit but for ours, or indeed for non-believers, as the case of the Apostle shows.[251]

For us it is clear that the sun stopped moving at the

command of God in the same sense that it appears to us that it moves. In the same way God is said to have made the moon to be a "great light"; but whoever says that this is true only relative to us does not, as Chrysostom says, thereby conceal the truth or deny God's action. Nor would that person have to agree with Epicurus or Lucretius, who maintain that the stars are exactly as large as they appear to us. Should we say, then, that the moon is a great light and agree with the interpretation of the impious Epicurus, who was not an astronomer, and thereby reject Chrysostom's explanation, which is based on astronomy and which agrees with common experience?

Our senses also tell us that a rainbow is formed when a cloud of dew receives and distorts the oblique rays of the sun, as is clear from physics.[252] But the divine Scriptures say in Genesis 9 [9:11–16] that every time it rains God will post this sign in the clouds as a memorial of himself and for our reassurance that he will never again destroy the earth with a deluge. If anyone, using this text, were to reject physics out of a fear that the rainbow would be attributed to the sun and not to God, he would be thought to be mad and ignorant of the Scriptures. For whatever nature does is the work of God; this is the law and command of God, as is proven by Chrysostom and Ambrose in his *Hexameron*. Therefore, since in the book of Joshua God is said to have stopped the sun even though it does not move, whoever says that this happened by arresting the motion of the earth does not deny the miracle but explains it, just as the physicist does not deny that God causes the rainbow but explains how he does it and what natural and reasonable means he uses.

REPLY TO THE SEVENTH ARGUMENT

It is easy to reply to the seventh argument. Deborah and Jude are speaking about the paths and motions of the planets, not the stars. Also the stars are not located in the heavens like knots in a board, as Aristotle thought, but they move by themselves, as the Master of the Sentences,[253] Augustine,[254] and all the Fathers thought. Also in Question 70, Article 1,[255] St. Thomas, following Chrysostom, agrees with Ptolemy rather than Aristotle that the stars are not like knots in a board, as all the Pythagoreans also said. In the same place Thomas points out that the planets are said to move in the lower, not the higher, part of the firmament.

In regard to Esdra we grant that the sun crosses the heavens in a circle. But in that text it is not said that the sun is moved but rather that it moves (which can be understood in the sense that the sun moves the planets in a circle by its light, as Pliny shows in Book II).[256] And the sun rotates the planets together with their spheres, or rather with the air and gas which compose a circle of vapors around the planets, as Galileo[257] and the Pythagoreans prove, and as Copernicus insinuates when he speaks of the great orb of the earth. But Chrysostom replied that it is enough to say that the sun rotates the stars to be able to say also that it rotates the heavens in which the stars are located, for he believed that the whole is fixed.

In fact the astronomers at the time of Augustine who proved with certitude, as he said,[258] that the heavens are fixed, were able to prove this only of the starry heavens and then only for the fixed stars, not the planets. The only possible way to prove mathematically that the heavens are immobile is in terms of the relations of the fixed stars and the planets to the earth, as Copernicus, Galileo, and the Pythagoreans have done. Hence with his phrase "with certitude" Augustine indicates that this was his opinion, although he was not an astronomer. But in the same place he also warns us not to assert that the contrary is a matter of faith, as we pointed out in the Second Hypothesis and in the reply to the second argument. See Augustine's commentary on Genesis, Book II, chapter 10.

In regard to the "speedy course of the sun" in Esdra, this is adequately understood as the sun's rotational motion on its own axis, which has been proven by Telesio[259] from sensory experience and by Galileo[260] from the translational motion of sunspots. Therefore the sun rotates on its axis and not around the earth. Or one could say, as above, that this is due to the appearances.

REPLY TO THE EIGHTH ARGUMENT

To the eighth argument I reply that it is by no means contrary to the Scriptures to say that there is water in the heavens. Rather it is a denial of this which would contradict Scripture and the Catholic faith. And as a result there must also be earth there. For water can be located only in a solid body like the earth and not in the tenuous and most thin heavens, nor in the stars whose heat would evaporate it into steam, even though some theologians who were worried about this have said that the water there is in the form of ice.

Furthermore Moses says in Genesis 1 [1:6–7], "The firmament divides the waters which are below it from the waters which are above it." And David in the Psalms [103:2–3] says, "He stretched out the heavens like a tent, which covered the waters above it," and elsewhere [Psalm 148:4–5], "The waters which are above the heavens praise the name of God." Daniel says the same thing in his song [Daniel 3:60], as does the whole of Scripture. Hence Origen,[261] who interprets "the waters above the heavens" to be angelic waters, disagrees with St. Basil.[262] And St. Augustine, who at one time adopted this interpretation in Book XIII of his *Confessiones*, abandoned it in his *Retractationes*, II, 6.

In Part I, Question 68,[263] St. Thomas presents three views about the nature of the firmament.

The first is the view of Empedocles[264] and other Pythagoreans that it is composed of the four elements. According to this view there is a clear and unproblematic explanation of the meaning of Scripture in the following two statements by Moses; namely, that water actually exists in the heavens and in the stars, and that the firmament was created on the second day, even though the heavens are said to have been made at the beginning, while the firmament made on the second day is also called the "heavens" by God.

The second is Plato's view that the firmament is composed of fire. (However I find in the *Timaeus*[265] a different view and account by Plato; namely, that it is composed of the four elements, even though Ficino[266] interprets this to refer to the basic qualities of the elements, i.e., the transparency of water, the solidity of earth, the mobility of air, and heat and light which are truly present in fire alone. St. Thomas did not read Plato himself, as he admits, because Plato had not yet been translated into Latin.)

But the text of Moses does not agree with this view on two points, as St. Thomas says. The first objection is that, according to Plato's view, the creation of the firmament is the same thing as the creation of fire. Hence the work of creation is the production of the elements, and is to be understood as occurring all at once when it was said, "In the beginning God created heaven and earth" [Genesis 1:1]. Now all the Fathers agree that in the subsequent days only an adornment of the world occurred. Thus a difficulty arises because if fire came into being on the second day, this was an act of adornment and not of creation. But this is nonsense if the two events are taken to be the same thing, for then the element of fire came into being after the element of fire was created. The second objection is that the "waters above the heavens" could not correctly be understood as true water, since a firmament of fire is not compatible with water.

Now both Basil and Chrysostom believe that the heavens are composed of fire, and so they answer these objections raised by St. Thomas. To the first objection Chrysostom says[267] that initially Moses speaks only generally when he says, "God created heaven and earth," and then later he speaks specifically about how they were created. And Basil replies[268] that the first day speaks about the immobile empyrean heaven, while the second day speaks about the starry heavens. To the second objection Basil replies that the waters above the firmament are frozen and have the role of dissipating the heat of the empyrean heaven and thus do not fall downwards. He also adds that the firmament could be understood to be dense air and that rain waters are above it, as in the phrase "the waters above the heavens." Nevertheless Basil and Ambrose and the Master of the Sentences[269] insist that these are truly water. This is also accepted by St. Bonaventure, Bede, and most of the Fathers, who maintain that the starry heavens were made from water, and that above them there is frozen water. In his *De Genesi ad litteram*, Book II, chapter 3, Augustine says that the starry heavens are fire and that the stars are made of fire. But he soon changed his view and said that the waters above the heavens were gradually raised up with the dense vapors above the air, which can thus be said to be the firmament. The Master of the Sentences reports this view without disagreement. And so the Fathers and the Scholastics have maneuvered in a variety of ways to save the text of Moses from the absurdities which arise if one says that the starry heavens are mobile and are not composed of the four elements.

The third view reported by St. Thomas is Aristotle's account that the heavens are composed of an unchangeable fifth element.[270] Thomas admits that this view is far less likely to save the text of Scripture. For the text says that the firmament was created on the second day from pre-existent matter, which in turn was created on the first day. Or rather, as St. Thomas says, it was created before there were any days at all, since it is said, "In the beginning God created heaven and earth, but the earth was formless and empty, and the spirit of God moved above the waters" [Genesis 1:1-2].

Augustine[271] understands "earth" and "water" here to refer to primary matter, which uneducated people could not understand unless it were described as corporeal, as St. Thomas also says in Question 68, Article 1.[272] In the same place Thomas argues that, according to Aristotle, the heavens are incorruptible by nature. Hence they have a matter which cannot accept another form; and thus it is impossible that the firmament was created on the second day. Furthermore it would be impossible to say that Moses placed real waters above the firmament, which contradicts what Scripture and the Fathers teach. Nor would it be true to say that waters from the lower world gradually ascended above the firmament.

In defense of Aristotle some say that the waters above the firmament are to be identified with the empyrean heaven, which is called water because it is similar to the transparency of water. For the same reason others call this the crystalline heaven, which they take to be immovable. Although St. Thomas sees that Aristotle's view can be defended only with difficulty, he still suggests a defense by saying in its favor that, if the firmament is composed of a fifth essence, then the firmament which divides the waters from the waters is primary matter, which according to Augustine is designated by the word "water."

But all these views, which were put forward to reconcile the opinions of Plato and Aristotle with Moses, are full of intractable difficulties, and they distort the text into a mystical meaning or into no meaning at all. But in his *De doctrina christiana*[273] Augustine teaches that when the literal meaning can be established, one should not take refuge in a mystical sense without first stating and explaining what that literal sense is. And in Book II of *De Genesi ad litteram*[274] he praises Basil's view[275] that the firmament is composed of air, because this is not contrary to the faith and can easily be believed on the basis of the text. In our *Questiones*[276] we have accepted that same opinion over the others because we do not have a better one.

Telesio approved of it because it is not inconsistent with his view that the whole heavens are composed of fire. But now, if Galileo's observations are true, I see even greater difficulties in the views discussed above.

First what Chrysostom says, namely, that what for Moses was created on the second day is a repeat of what was created before any of the days, is not only contrary to many of the Fathers but also does not seem likely. For all the Fathers agree that creation occurred before any of the days, and adornment is what happened during the days, as Augustine,[277] Thomas, and the Master of the Sentences[278] teach. Not everyone agrees with Basil's view that the unchanging empyrean heaven was created before the days and the starry heaven on the second day, for he states this with no basis in the Scriptures but only to defend his own view.

Further, in Book X of *De civitate Dei*[279] Augustine, appealing to Porphyry,[280] says that the empyrean heaven is the same as the starry heaven, and he says that it is called empyrean because it is made of fire. Indeed the stars are made of fire, and Wisdom 18 [17:5] speaks of the "bright flames of the stars." According to Augustine it is called the aether because it is made of flames and not because of the speed of its motion, as Aristotle thought. Thus the empyrean heaven is made of fire. But fire is always in motion, as we have proven in our *Questiones physiologicae*,[281] and it is extinguished when it is deprived of motion, as Averroës says. Further the heavens move in a circle because they are made of fire and cannot stop, as is said by Plotinus[282] in *Ennead* II and by Xenocrates[283] and Porphyry, responding to Aristotle, who believed that fire is naturally at rest in its own sphere but is moved contrary to its nature in a circle by the heavens. Therefore it is not correct to say that the empyrean heaven or fire is different from the starry heavens, which clearly surpass all in their heat and light.

We simply do not know whether that other heaven located above the starry heavens emits light or heat, for its

light is imaginary and does not come across to us. As a result Basil decided that he could not defend this view, and
thus he thought that the firmament was composed of air.
This is a good way to avoid the objection, but it does not
explain the text of Moses. For above the air there is no
water in the form of clouds, except the vapors which rise
from the earth. But vapor is not water, even though water
can be produced from it. But water is also generated from
air, according to St. Thomas and Aristotle, the latter of
whom thought that subterranean fountains are generated
and replenished from the air. Therefore the water which
Moses located above the firmament is not truly water but
is matter from which water can be generated as much as
air or fire, according to St. Thomas and Aristotle. Hence
this matter does not divide the waters from the waters.
Rather it divides any firmament from any other firmament.
Thus one can imagine whatever one wishes and can abuse
words by saying that fire is wood or that water is vapor.
And this generation will not be a transformation but a separation, as the ancients said, but which Aristotle and St.
Thomas condemn. On their opinion see my *Questiones*
and my *Metaphysica*.[284]

Furthermore watery vapors are not above the dense
air but within it, and therefore they do not divide the waters from the waters. The same difficulties are encountered
by those who think that the heavens are diaphanous and
thus similar to water and to crystal. But this similarity is
not the reality which the word "water" signifies. For both
air and the sun are diaphanous, but they are not thereby
water. Therefore it seems that these are fictional views
which do not arise from the authority of Scripture but
from the difficulties we run into when we try to explain
Moses not according to the philosophy of Pythagoras but
according to the philosophy of Aristotle and Plotinus. For
Plato agreed with Pythagoras, but Plotinus thought that
the heavens by nature were composed only of fire.

The solution that the true waters are really ice has no

value, unless we grant that the stars are worlds in which water is naturally present, as on the earth, and that the stars reflect light because of water and air, such that they appear as stars to the inhabitants of neighboring stars, just as the moon appears to us. But it would be extraordinary and contrary to natural law for ice to be mixed together with fire without being liquified and dissolved by the fire. And Augustine says that there is no miracle involved in this case. Basil marveled at this but thought that a better explanation could not be found.

The arguments used to save the text according to the views of Aristotle are even more absurd, as St. Thomas admits. He thought that the first objection against Aristotle was unanswerable, given that the six days are natural days and not angelic days, as was thought by all the Fathers, Chrysostom, Bede, Jerome, Origen, Gregory, Ambrose, Basil, Procopius, and all the others except Augustine, who says that Aristotle agrees with Moses, but only with difficulty, if the days are angelic. But to defend Aristotle should we accuse all the Fathers of ignorance and impiety for having accepted an impossible, absurd, and false interpretation? And should we place Aristotle above the holy doctors in the teaching authority of Christ? We must reject such wickedness and foolish blindness.

Moreover for Aristotle it would be completely impossible for there to be water above the heavens, unless one were to speak equivocally of "spiritual water." But this was rejected by all the Fathers, for if we wish to be Christians and not Aristotelians, then there must be physical water located there. This view is supported by the Jewish rabbis and the Pythagorean philosophers, as well as by Basil, Ambrose, and other Fathers and the Scholastics. Furthermore St. Thomas[285] rejected both of these views and belittled their explanatory power. For he said that it is impossible for vapors from the earth to rise above the starry heavens, which are composed either of fire or the fifth essence. This would be even more impossible if the heavens are composed of the four elements, both because of the distance involved and because the water would be changed into a different kind of substance. Also there do not exist any minute particles of water, except according to Anaxagoras and Empedocles, which we have discussed elsewhere.[286] But if we decide to accept their view, then the vapors could rise up from the stars themselves, if they are worlds composed of the four elements as our world is.

Moreover what St. Thomas adds from Augustine in defense of Aristotle is not satisfactory. For Moses did not divide primary matter with the firmament but real water, as all the Fathers say. Also what David says would seem to be absurd: "You stretched it out with the waters above it" [Psalm 103:3], which would mean that he stretched out the heavens with primary matter. And elsewhere he says, "The waters above the heavens praise the name of God" [Psalm 148:4]. But how could a thing which is formless and nearly nothing praise God, unless it either actually had reason like the aqueous angel of Origen,[287] or it was adorned with very great beauty which can be said to display the praise of God in analogy to a person, as Basil says?[288] Also what marvels would Moses have related and taught if by the division of the waters from the waters by the firmament he had meant the division of matter from matter? It is clear that the waters under the firmament are truly water, like our seas; hence those above the firmament should likewise be true waters. For a distinction between things of the same kind is not a distinction between things having nothing in common.[289] Therefore although St. Thomas is inclined towards the view of Empedocles because it clarifies the meaning of the Christian Scripture, it seems that in his modesty he piles up all these interpretations to satisfy the doctors and the philosophers and to show that the Scriptures abound with many meanings.

On the other hand, it is clear that not only Moses but also Solomon and experience agree with Empedocles and Galileo. Solomon says in Proverbs 8 [8:27–28], "When he established the aether on high, and encircled the abysses with a fixed law, and balanced the fountains of the waters." According to Augustine and Porphyry, the starry heavens are called the aether because they are made of flames.[290] And the "abysses" are immense waters which can be imagined as fluid and as contained within many valleys in many worlds and surrounded by earthen walls, like our sea on our earth. The fountains of water are said to be balanced perhaps because they are in the stars, for every world has a proper balance at its center.

I know that there are other explanations, but they all antedate the discovery of new phenomena. It is now known that small clouds rotate around the sun, which cannot have risen up there from our earth, as both St. Thomas and reason itself teach. Likewise in 1572 a new star with no parallax was formed from vapors in the constellation of Cassiopeia, as has been reported by Tycho[291] and numerous other astronomers who have seen and observed it. Therefore there are vapors on the stars. Also instruments have proven that comets are located above the moon, which Aristotle denied. But vapors cannot rise up there from our earth. Therefore on the stars there are earth and water, especially since certain comets are observed to stay close to certain stars.

Also minute drops of water, as Ambrose and Augustine have said, cannot have risen there from our earth; hence they come from the water which is on the stars. Further, according to the Apostle Peter, the heavens "will be dissolved by heat" [2 Peter 3:10], and according to David, "they will grow old and perish" [Psalm 101:27]. But if the heavens are composed of fire or of the fifth essence, this cannot be understood without distorting the Scriptures. St. Clement, Hilary,[292] and Catharinus[293] understand this to refer to the higher heavens and not to the air, as Augustine said; therefore, etc. Also Galileo has shown that there are mountains on the moon, and Scripture agrees with him because Genesis 49 [49:25] and Deuteronomy 33

[33:13–15] speak of fruits and mountains and hills in the heavenly bodies.

Therefore sacred Scripture, taken in its literal sense in all its passages, agrees only with Empedocles, and it agrees with the others only arbitrarily or in a mystical sense. But Empedocles was a Pythagorean, as also is Galileo. So Galileo ought to be praised. For after so many centuries he has protected the Scriptures from ridicule and distortion. And he has shown that the wise men of this world are fools, and Sacred Scripture should not be adapted to them, as has happened up to now, but they ought to be adjusted to fit the Scriptures. And this does not degrade our planet. For humans will be raised up with Christ, their leader, above the stars and above all the heavens. So from this it is clear that we are better off than those wise men.

Reply to the Ninth Argument

I respond to the ninth argument by denying the conclusion. Galileo has not said that there are many worlds but rather that all the celestial systems constitute one world under one immense heaven. The theologians who follow St. Basil and St. Clement claim that there are three worlds: the world of the four elements, the world of the heavens, and the spiritual world above the heavens. Moses' three-part tabernacle was constructed as an analogy to this, according to Philo, Josephus, Clement of Alexandria,[294] Jerome, and Sixtus of Siena.

But Galileo has nothing to say about these theological matters. Rather with his marvelous instruments he has recently shown us previously hidden stars, and he has taught us that the planets are similar to the moon, that they receive their light from the sun,[295] that they rotate around each other, that transmutations of the elements occur in the heavens, and that there are vapors and clouds and many celestial systems in the orbits of the stars.

Since we can touch these with our hands, as it were, we know that Moses spoke truly of the "heavens of the heavens" [Deuteronomy 10:14] and of the water and mountains and other things in the stars. And we know that we can understand the Scriptures literally and not with effort and distortion and imaginary fictions. And thus we are protected from the deceptions of the philosophers who, not believing these things, have taken refuge in mystical meanings, just as Persian heretics[296] today have introduced impossible fictions about celestial and divine things in their explanations of Mohammed.

Furthermore it must be noted that nowhere in the canons of the Church is there to be found a decree which denies that there are many worlds. Nor does St. Thomas say that this is contrary to the faith when he discusses the matter in Part I, Question 47, Article 3.[297] The passage in John [1:10], "The world was made by him," does not deny that God made other worlds at other times; it states only that he made our world. But St. Thomas correctly shows that it would be an error of faith to claim that there are many worlds without any order as a whole, as Democritus[298] and Epicurus[299] thought. For from this it follows that these worlds came into being by chance, as they believed, without being ordered by God.

But to say that there are many lesser worlds which are ordered into one whole by God is not contrary to Scripture, but only to Aristotle. St. Thomas' argument (that there cannot be another earth in another world outside of this one because that earth would be attracted to this one and would abandon its place) is based on Book I of Aristotle's *De caelo*[300] and is not valid. For my heart does not go to where your heart is. All things are stable at their own center, and they preserve and rejoice in the specific similarity of their parts. What belongs to the moon is attracted to the center of the moon, and what belongs to Mercury is attracted to the center of Mercury; outside of their own sphere they do not detect anything better for themselves. And if the stars all have the same nature, as the Peripatetics claim, then why are they not attracted to each other and the parts of one to the parts of another?

Moreover when the University of Paris published a list of articles from the writings of St. Thomas which needed correction, the proposition, "There cannot be another world,"[301] was included in the list. The objection was that God cannot be limited by theologians, whose role is to speak of divine rather than natural matters. But in truth St. Thomas did not limit the power of God as is claimed here (even though he did not explain himself as clearly as he could have), for elsewhere he states the opposite, i.e., that God can make many worlds and many earths. Rather the criticized idea came from the traditional understanding of Aristotle, as Cajetan also has noted.

In his commentary on Book I of the *De caelo*[302] where Aristotle discusses this topic, St. Thomas teaches that the plurality of worlds is not contrary to the faith, but to Aristotle. Likewise from what Galileo has said it does not follow that there are many human species and that Christ also died elsewhere. The conclusion is not valid in either case. Granting that Christ did not die in the other hemisphere, many theologians including Augustine have concluded that that hemisphere does not exist. But today this argument has been refuted by experience. If there are humans living on other stars, they would not be infected by the sin of Adam since they are not his descendants. Hence they would not be in need of redemption, unless they suffered from another sin. This provides an interpretation of Ephesians 1 [1:7–10] and Colossians 1 [1:20], "He reconciles with his blood what is in the heavens and what is on the earth." But we do not know this, and thus we stay with the old interpretation of the Fathers.

In fact in his *Letters on Sunspots*[303] Galileo explicity denied that there can be humans on other stars (which we have proven with a physical argument in our *Questiones*).[304] However there could be beings there of a different nature who are similar to us but not the same as us, no matter what Kepler has said about this in a playful and joking way and as a pure hypothesis in his *Dissertatio*.[305]

Furthermore even if the theory of a plurality of worlds is false, this does not affect Galileo, who, using his senses and not his imagination, detected not a plurality of worlds but a plurality of systems in this world which are ordered in a unified way. This destroys Aristotle's argument in *Metaphysics*, XII,[306] that there are many prime movers. Before Galileo the same thing was said by Cardinal Cusa, Kepler, Bruno, and others.

The fact that the divine Scripture does not say something does not mean that it is not true; an argument from a negative authority is a logical fallacy. Scripture never speaks of the other hemisphere. Should we therefore agree with those atheists who use such an argument to condemn Moses because he did not speak of the antipodes, even though on their side they have Augustine, who did deny the antipodes precisely because Moses did not speak of them? Indeed with this kind of reasoning we should say with Luther that Peter was never in Rome because Luke makes no mention of this in Acts. Let us bid farewell to such ignorant and senseless trifles. Paracelsus' foolish opinion is so far removed from Galileo that we will not even stop to consider it. Moses was silent about all these things because he wrote about the law given to our world, not about the systematic physics of the whole universe, and indeed he said of our world only what was needed for the law.

REPLY TO THE TENTH ARGUMENT

In reply to the tenth argument I deny that Galileo has given rise to an active scandal, the only thing expressly prohibited in the Gospel. For he has not called on us to do anything wrong but rather to search for the truth, which God has entrusted to us and ordered us to do, as is clear from the Second Hypothesis in chapter 3 of this book and from the Gospel itself, where Christ warns us under the greatest penalties not to bury the talents he has given us.[307] In his commentary on Ezekiel, St. Gregory says, "If a scandal arises from the truth, it is better to let the scandal occur than to abandon the truth."[308] And when the Pharisees were scandalized by what Christ said, he replied in Matthew 15 [15:14], "Let them be. They are the blind leading the blind."

It is an error, if indeed not a heresy, to say that the views of Aristotle about the heavens and the structure of the world have been accepted by the Scholastics as consistent with theology and in need of no further investigation. In fact all the Fathers disagree with Aristotelianism about the heavens and the structure of the world, as I have shown above in the Second Hypothesis and in my replies to the first and eighth arguments. And the Scholastics, whose leaders are Peter Lombard and St. Thomas Aquinas, clearly declare that on these matters the theories of Aristotle are irreconcilable with the teachings of Moses and the Fathers (as is clear from my replies to the second and eighth arguments and from what I have already said above in this article.)

But the discoveries of Galileo agree with the sacred Scriptures and save them from the distortions of the theologians and the ridicule of the philosophers. He has shown that the philosophers are wrong, and that the testimony of the Fathers is truer than that of the philosophers. I do not know why anyone would choose to be blinded by a rage of false passion without any understanding or to attack cold-bloodedly a teaching based on sensory facts.

REPLY TO THE ELEVENTH ARGUMENT

The eleventh argument has already been answered in the first assertion of the Second Hypothesis and in its corollary. It was shown there that God is pleased when we study his book and that it is not vain to investigate the heavens. Rather this is useful to reveal the glory of God and to increase our faith in the immortality and divinity of the human soul.

Also Cato's verses are not comparable to the contrary and noble verses of Ovid. Furthermore David recommends that type of study as being sublime and not prideful. Let us add that Cato's argument, "Since you are mortal, attend to the things which are mortal," is contrary to the faith, since it must apply not only to the body but also to the soul. Since the soul is immortal and capable of being divine, the investigation of divine things cannot be foreign to it. Thus David says [Psalm 68:32], "Seek God and your soul will live," and elsewhere [Psalm 104:4], "Seek his face always," and concerning heavenly things [Psalm 18:1], "The heavens declare, etc.," and [Psalm 8:3], "I will see your heavens, etc.," and elsewhere [Psalm 118:129], "Your works are wonderful, and thus my soul examines them." From the second and third assertions it is clear that the Fathers placed no limits on these investigations. And from the fourth assertion it is clear that whoever forbids inquiry commits an error and that we should pursue knowledge with moderation. These matters have been treated fully in our *Theologia.*

CHAPTER 5

Evaluation of the Arguments in Galileo's Favor Stated in Chapter 2

I think that it would be very difficult today to refute all the arguments which have been proposed in Galileo's favor. For many years I believed that the heavens were made of fire, that these heavens were the source of all fire, and that the stars were made of fire, as was also thought by Augustine, Basil, other Fathers, and more recently by my compatriot Telesio. In my *Questiones* and *Metaphysica* I tried to refute all the arguments of Copernicus and Pythagoras.[309]

But I came to doubt that all the stars are composed of fire after the observations made by Tycho and Galileo, which have confirmed that small clouds rotate around the sun, that a new star has come into being, and that comets have occurred beyond the moon in the starry heavens. This doubt was strengthened by the changing phases of the moon and Venus and by the large spots on the moon and Jupiter. Although I had given an answer to the objection that the starry spheres would have to cross through immense distances in a single moment, this still made me uncomfortable. The stars located above and rotating around Jupiter and Saturn did not support the view that there is only one sun and one center of attraction, i.e., the sun, and another center of repulsion, i.e., the earth, as we have said in our *Physiologia*.[310] On the other hand, the notion that the heat of the fixed stars is the same as that of the various planets raised doubt about the opinion of Galileo and others that there are many suns.

As a result I will suspend my own judgment on these matters as I respond to the arguments in Galileo's favor. And I am prepared to obey the commands of the Church and the judgment of others who know more than I do.

REPLY TO THE FIRST, SECOND, THIRD, FOURTH, FIFTH, AND SEVENTH ARGUMENTS

The reply to arguments 1, 2, 3, 4, 5, and 7 is the same. The theory of Copernicus and Galileo, which has been approved by many theologians, is probable but not certainly true. This view has not been defined by a General Council nor by the personal action taken by Pope Paul III, assisted by the Holy Spirit, who merely permitted Copernicus' book stating this theory to be published as not contrary to the faith. When a pope approves a teaching of holy theologians, he does not thereby establish it as a matter of faith but only as useful and worthy of study, as the doctors in Paris said in the articles they listed concerning St. Thomas. Otherwise Pope Gelasius[311] would have authorized the errors of Cyprian, Jerome, and others, when he received and approved them in Distinction 15 of the chapter *Sancta Romana*.[312]

As a result it is probable, but not absolutely certain, that there is nothing contrary to the Scriptures in these teachings, which have been permitted by the authority of the pope and theologians mentioned above. Modern theologians might see something which their predecessors missed if they more carefully and more cleverly study the heavens and the Scriptures, as I have proposed in this book that they do, or if they receive a new revelation.

For myself I must say that I do not see that anything harmful to the authority of the Sacred Scriptures arises out of Galileo's teachings. Indeed I think rather they are a benefit, as is clear from what has been said.

REPLY TO THE EIGHTH, NINTH, AND TENTH ARGUMENTS[313]

I do not know whether arguments 8, 9, and 10 really are in favor of Galileo, for theologians avoid mystical meanings and equivocal uses of "heavens," as is clear in St. Thomas, Part I.[314] But they certainly are arguments against Aristotle. In the supporting arguments for Galileo we have examined everything said by the theologians, and from this we have concluded that the Scriptures do not favor Galileo any less than those who speak out for the teachings of other philosophers, although I am prepared to believe anyone who knows more than I do about this.

In my *Metaphysica*, Part III, Book I, I have examined all the teachings of Galileo, Copernicus, and the Pythagoreans and have answered them as best I could. Likewise in the whole of my *Questiones physiologicae* I have treated at greater length the natural arguments against Copernicus. The physicist can read those arguments; here we are treating the topic theologically. But it is up to the Church to judge whether Galileo should be permitted to write about and debate these matters.

REPLY TO THE SIXTH ARGUMENT

I do not know what I would deny in the sixth argument. Count Pico della Mirandola claims that it is a true story that Aristotle once either read or heard recited the text of Moses about God freely creating the world at the beginning of time, and that he rejected it as boorish and unproven. Eusebius reports[315] that Porphyry said the same thing.

Ambrose testifies that Pythagoras was a Jew, but I do not recall whether it was in his *Sermones* or his *Epistolae*,[316] and I do not now have his books at hand. But I do remember that he said this. Indeed one of the commentators on Ambrose wondered how Pythagoras could be a Jew when others say that he was from Samos in Greece (Gabriele Barrio[317] of Francica says he moved from Samos to Calabria, which was part of Magna Graecia at the time), and answered that Ambrose would not have said this without clear historical evidence. This is justified not only by Ambrose's sanctity and reputation but also by arguments. For Pythagoras taught that we should abstain from certain foods, that there is only one God even though he said that the angels are second gods, and that all things are numbers (as did Moses in the construction of the tabernacle and Solomon who said [Wisdom 11:21] all things were created "in number, weight, and measure"). Also Pythagoras rivaled Moses as a lawgiver; on this point see my *Metaphysica*.[318] But all these things were also well known to the Jews.

It is likely that Pythagoras was born on Samos to a Jewish family, just as the Spartans are said in the books of the Maccabees [1 Maccabees 12:21] to be descendants of Abraham, for the Jews were dispersed into many parts of the world at the time of Abraham, Moses, and the Judges. According to Diogenes Laertius,[319] Plutarch,[320] Aristotle, and Galen,[321] Pythagoras was the first to promulgate to the gentiles this marvelous philosophy that the earth moves, that there are many systems in the heavens, that the sun is at the center, that the moon is another earth, and that all four of the elements, and not just water, exist in the stars. Therefore it seems that Pythagoras derived these teachings from Moses, for he could not have had such wisdom without a previous revelation.

Copernicus, who was struck by the observations of Francesco Maria,[322] began to consider these matters by studying the prior teachings of the Pythagoreans. My compatriot Timaeus[323] of Locri, a disciple of Pythagoras, demonstrated mathematically the diurnal motion of the earth; Philolaus of Croton proved its annual motion; and Copernicus seems to have added the motion of libration[324] (as I have shown in my *Questiones physiologicae*),[325] following the example of Thebit of Babylon and King Alfonso of Spain. In his commentary on *Metaphysics*, XII,[326] St. Thomas uses Simplicius' account to indicate which of these motions are necessary.

Even if Pythagoras were not a Jew and thus did not learn these things from the philosophy of his ancestors, still we know from history that he visited with the priests of Egypt with Pherecydes of Siros,[327] and with the Jews in Syria, in Egypt, and in that part of Judea which is between Syria and Egypt. Thus it was from them that he heard the law and the philosophy which locates water and mountains and earths in the heavens, and mountains on the moon, and similar things, which we have shown are found in the Holy Bible.

But as Aristotle scorned Moses the Jew, he also scorned Pythagoras, who was either a Jew or was influenced by Jewish ideas. Therefore we Christians, who spiritually are Jews as the Apostle says [1 Corinthians 9:20], are the ones to rescue the sacred philosophy of Moses from the insults of the pagans by using the most discriminating instruments and arguments. Why then do we murmur, as the Jews once murmured against Moses, who rescued them from the persecution of the Egyptians? The old rabbis, whose books I do not now have, teach almost the same thing.

Moreover Mohammed was a descendant of Ishmael and was educated by the Jews, whom he took back with him, as we know from much historical testimony gathered by the "Ecstatic Doctor" Dionysius Carthusianus in his books against Mohammed. In his *Dialogus cum Abdia Iudeo* and in the *Koran* Mohammed locates many seas and mountains and airy spaces in the heavens, and under the earth he places seven other earths and an ox holding them up. He seems to have derived these things from the Jews and the Talmud scholars (as is indicated by Sixtus of Siena, by Dionysius Carthusianus,[328] by others, and by the doctrines themselves). But it would be quite ignorant to think that whatever crosses the lips is inspired, as is clear from his own replies.

Mohammed confuses the true with the false here, as he does also in his histories of Joseph, David, Solomon, and our Lord Jesus Christ. Also he does not know how to distinguish between literal and metaphorical meanings. So he accepts as real the columns which hold up the world, and the rivers of cream and wine[329] mentioned in Job [9:6; 26:11; 20:17], and other such things. And he believes that to prevent the heavens from falling down, they are held up by Mount Caf, from which they acquire their greenish color. He derived this from some early Christians who, in their ignorance of geography, located the terrestrial paradise on top of that mountain, which they thought rose up into the heavens. This caused Anastasius Sinaita to ask how humans descended from there. I will omit what Bede says about these matters except for one point, namely, that it was from the rabbis that Mohammed derived the notion that there are many earths and seas and world systems above our heavens.

In conclusion, the views of Galileo and Empedocles, who derived his teachings from the Pythagoreans more than from any other philosophers, as St. Thomas says, are in agreement with the ancient and the modern interpretations of Sacred Scripture. Therefore they are also in agreement with the Sacred Scriptures themselves on astronomical matters. Moreover the Pythagoreans derived their teachings from the Jews, and Galileo now strongly agrees with these teachings, being motivated not by a trivial opinion but by sensory observations.

In my judgment, in agreement with what St. Thomas and St. Augustine have taught us in our Second Hypothesis, it is not possible to prohibit Galileo's investigations and to suppress his writings without causing either damaging mockery of the Scriptures, or a strong suspicion that we reject the Scriptures along with the heretics, or the impression that we detest great minds (especially since in our day the heretics disagree with everything said by Roman theologians, as Bellarmine has pointed out).[330] It is also my judgment that such a condemnation would cause our enemies to embrace and to honor this view more avidly.

In what I have written and will write in the future, I submit myself as always to the censure of our Holy Mother Roman Church and to the judgment of those who are wiser than I.

Our farewell greetings to the Most Illustrious Cardinal Caetani, the patron of Italian virtues.

NOTES

INTRODUCTION

1. See *Defense*, chapter 5, "Reply to the Sixth Argument."

2. Despite his considerable accomplishments as a poet, Campanella's Latin prose style is quite awkward and devoid of literary flourish; his Italian prose style is mediocre at best. According to Léon Blanchet (*Campanella* [Paris: Alcan, 1920], 18) this is due primarily to the inferiority of his early literary education.

3. As translated by Bernardino M. Bonansea in his *Tommaso Campanella: Renaissance Pioneer of Modern Thought* (Washington, D.C.: Catholic University of America Press, 1969), 33.

4. Telesio maintained a tripartite theory of the human person: (1) the body, (2) the spirit (very subtle warm matter centered in the brain, whose motions account for sensation), and (3) the immaterial mind, which informs both body and spirit, and whose tendency toward union with God demonstrates its immortality. Telesio did not claim to have developed more than a natural philosophy, and so many metaphysical and theological questions are simply left unexamined in his *De rerum natura*.

5. See *Defense*, chapter 3, "Proof of the Fifth Assertion," and note 174 to the translation.

6. For example, see *Defense*, chapter 3, "Proof of the Fourth Assertion."

7. Blanchet (*Campanella*, 21–24) has argued that Campanella's involvement in magic and the occult was the central problem in his clash with the authorities and that his commitment to Telesian philosophy was merely a means to that end.

8. Luigi Firpo, ed., *Apologia di Galileo* (Turin: UTET, 1968), 9. Firpo also argues that during this year Campanella was heavily involved in experimental work in science, which his later imprisonment made impossible to continue. Firpo's Introduction, "Campanella e Galileo" in the above edition, is a detailed study of the infrequent but significant correspondence between the two men, in which Galileo usually wrote to Campanella through a third party.

9. The best study of Campanella's imprisonment, trials, and torture is Luigi Firpo's *Il supplizio di Tommaso Campanella: Narrazioni, Documenti, Verbali delle torture* (Rome: Salerno Editrice, 1985).

10. If he had confessed, then, since he previously had been convicted of heresy, he would have thereby become a "relapsed" heretic, which automatically meant execution.

11. *Le Opere di Galileo Galilei*, Edizione Nazionale, edited by A. Favaro (Florence: G. Barbèra, 1890–1909, 1929–39, 1964–66), XI, 21–26.

12. For a detailed discussion of this episode, see Blanchet, *Campanella*, 345–55.

13. "Utrum ratio philosophandi, quam Galilaeus celebrat, faveat sacris Scripturis, an vero adversetur."

14. The solar or tropical year is the period between two successive vernal equinoxes as the circle of the sun's apparent daily motion slowly drifts north and south of the earth's equator.

15. The calendar reform did not occur until the introduction of the Gregorian calendar in 1582 by Pope Gregory XIII. Ten days (October 5–14) were dropped from that year, and to avoid a recurrence of the same problem, three leap days (February 29) are dropped every 400 years in the century years not divisible by 400. Thus there is a February 29 in the years 1600 and 2000, but not in the years 1700, 1800, and 1900. This calendar reform was accepted in Europe and elsewhere only quite slowly because of the religious rivalries after the Reformation. For an excellent analysis of the whole story of this calendar reform, see G. V. Coyne, S.J., M. A. Hoskin, and O. Pedersen, eds., *Gregorian Reform of the Calendar: Proceedings of the Vatican Conference to Commemorate Its 400th Anniversary, 1582–1982* (Vatican City: Pontificia Academia Scientiarum, Specola Vaticana, 1983).

16. Although the Council of Trent began just two years after the publication of Copernicus' book, the latter was of no concern to the Council in any way. The Council dealt rather with reforms needed in the Church and with the challenges it faced from the Protestant Reformation, which had begun with Martin Luther in 1517.

17. See R. J. Blackwell, *Galileo, Bellarmine, and the Bible* (Notre Dame, Ind.: University of Notre Dame Press, 1991), 181–84, for my translation of the entire decree.

18. See Bellarmine's Letter to Foscarini, 12 April 1615, in Galileo, *Opere*, XII, 171–72. For a fuller account of the evolution of views on biblical exegesis from Trent to Bellarmine and Galileo, see R. J. Blackwell, *Galileo, Bellarmine, and the Bible*, chapters 1–3.

19. Bellarmine's Letter to Foscarini, 12 April 1615.

20. These developments have been studied for generations in exhaustive detail and with a wide range of interpretations. For some recent studies, see R. J. Blackwell, *Galileo, Bellarmine, and the Bible* (1991); J. Langford, *Galileo, Science, and the Church* (New York: Desclee, 1966); L. Geymonat, *Galileo Galilei* (New York: McGraw-Hill, 1965); P. Redondi, *Galileo Heretic* (Princeton, N.J.: Princeton University Press, 1987); Giorgio de Santillana, *The Crime of Galileo* (Chicago: University of Chicago Press, 1955).

21. Galileo, *Opere*, XIX, 322–23. The translation is from R. J. Blackwell, *Galileo, Bellarmine, and the Bible*, 122.

22. For the documents relating to Galileo's trial and to the events leading up to it, see S. M. Pagano, ed., *I documenti del processo di Galileo Galilei* (Vatican City: Pontifical Academy of Sciences, 1984) and Maurice A. Finocchiaro, ed., *The Galileo Affair: A Documentary History* (Berkeley: University of California Press, 1989).

23. L. Firpo, *Apologia di Galileo*, 19. Firpo's interpretation has been accepted by Antonio Corsano ("Campanella e Galileo," *Giornale critico della filosofia italiana* 44 [1965]: 318), and was anticipated by Luigi Amabile, *Fra Tommaso Campanella: La sua congiura, i suoi processi, e la sua pazzia* (Naples: Morano, 1882), I, 183–86.

24. Pietro Iacopo Failla's Letter to Galileo, 6 September 1616. Galileo, *Opere*, XII, 277.

25. Firpo, *Apologia di Galileo*, 21.

26. The Latin text of this sentence is: "Tu vide quid recte dictum sit; quid item defendendum tibi aut renuendum; quando a sancto senatu id in mandatis habes."

27. The decision that Cardinal Caetani would make these corrections must have been made less than "ten days" after the condemnation, since Galileo already knew about it by the next day. See Galileo's Letter to Curzia Picchena, 6 March 1616, in *Opere*, XII, 244. However that point is not relevant to Firpo's argument here.

28. Galileo, *Opere*, XII, 106–7.

29. See Salvatore Femiano, ed., *Apologia per Galileo* (Milan: Marzorate Editore, 1971), 21–30, where all the relevant textual evidence and arguments on both sides are cited and evaluated. The fullest discussion of the issue in English is in Bernardino M. Bonansea, "Campanella's Defense of Galileo," in *Reinterpreting Galileo*, ed. by William A. Wallace (Washington, D.C.: Catholic University of America Press, 1986), 206–14.

30. Tommaso Campanella, *Lettere*, ed. Vincenzo Stampanato (Bari: Laterza, 1927), 223.

31. Not published until 1638 as Part I of his *Philosophia realis*.

32. Femiano, *Apologia per Galileo*, 27.

33. Chapter 5, "Reply to the Eighth, Ninth, and Tenth Arguments."

34. Additional discussions in English of the content of the *Defense* can be found in B. Bonansea, "Campanella's Defense of Galileo," 214–39; in J. Langford, "Campanella on Scientific Freedom," *Reality* 11 (1963): 133–50; and in J. Langford, "Science, Theology, and Freedom: A New Look at the Galileo Case," in *On Freedom*, ed. Leroy S. Rouner (Notre Dame, Ind.: University of Notre Dame Press, 1989), 108–25.

35. Chapter 3, "First Hypothesis."

36. In his Letter to Foscarini, 12 April 1615, Bellarmine also points out that the new astronomy has not been proven, but then goes on to advise that if and when it ever were to be proven, theologians should then reinterpret the relevant passages of Scripture accordingly, or at least say that we do not fully understand them.

37. For an excellent discussion of Galileo's dependence on patronage from the Tuscan royal family, see Richard S. Westfall, *Essays on the Trial of Galileo* (Vatican City: Vatican Observatory Publications, 1989).

38. For an English translation of both of these letters, see M. Finocchiaro, ed., *The Galileo Affair*, 49–54 and 87–118.

39. For a discussion of these developments, see R. J. Blackwell, *Galileo, Bellarmine, and the Bible*, chapters 3 and 5.

40. For a complete English translation of Foscarini's *Letter* and an analysis of the controversy surrounding it, see R. J. Blackwell, *Galileo, Bellarmine, and the Bible*, appendix VI and chapter 4.

NOTES TO THE TRANSLATION

1. Campanella seems to have had a special liking for this metaphor, since it appears in several of his writings. This may explain why Tobias Adami, the editor of the 1622 edition of the *Apologia pro Galileo*, used this metaphor in the first sentence of his Preface. For a discussion of the earlier use of this metaphor in the sixteenth century, see Carlo Ginzburg, *Il formaggio e i vermi* (Giulio Einaudi Editore, 1976); English translation by John and Anne Tedeschi: *The Cheese and the Worms* (Baltimore: Johns Hopkins University Press, 1980; New York: Penguin Books, 1982).

2. The telescope was invented by Hans Lippershey in Holland in 1608, and was considerably improved by Galileo in 1609. At first the "optical tube" was referred to by various names. The word "telescope" was coined in 1612 by Prince Federico Cesi (1585–1630), a close friend of Galileo, and the founder and guiding spirit behind the Academy of the Lynx, the earliest scientific society in Italy. On the coining of the name, see Galileo Galilei, *Le Opere de Galileo Galilei*, Edizione Nazionale, edited by A. Favaro, 20 vols. (Florence: G. Barbèra, 1890–1909; reprinted: 1929–39; 1964–66), XI, 420.

3. In the sixteenth and seventeenth centuries Pythagoras and his disciples (we really do not know which of their ideas originated with which individuals of that group) were regularly credited with holding a heliocentric astronomy. But actually they held that both the earth and the sun revolve around the center of the universe, where is located the Central Fire or the Hearth of Zeus, which was invisible to them because it always faces the other side of the earth.

4. Foscarini's (c. 1580–1616) book was entitled *Lettera sopra l'opinione de' Pittagorici e del Copernico...* (for an English translation see R. J. Blackwell, *Galileo, Bellarmine, and the Bible* [Notre Dame, Ind.: University of Notre Dame Press, 1991], 217–51). In his book Foscarini attempts a reinterpretation of Scripture from the perspective of Copernican heliocentric astronomy. The book was condemned and placed on the

Index librorum prohibitorum by a Decree of the Congregation of the Index, dated 5 March 1616, which also condemned Copernicanism as "false and completely contrary to the divine Scriptures."

5. Sebastiano Fantoni, O. Carm. (1550–1623), joined the Carmelite Order in 1567 and was General of the order from 1612 until his death.

6. Nicholas of Cues (Nicolaus Cusanus, de Cusa) (1401–1464), appointed a cardinal in 1448, argued in his *De docta ignorantia* (1440) that there are multiple worlds and that the earth is neither at rest nor in the center of the universe, views which he advocated not from any work in astronomy but from his philosophical analysis of the notion of an infinite universe.

7. Nicholas Copernicus (1473–1543) was the author of *De revolutionibus orbium coelestium* (1543), which opened the debate against the geocentrism of Aristotle and Ptolemy.

8. Georg Joachim von Lauchen, called "Rheticus" (1514–1576), was a close friend and student of Copernicus. His *Narratio prima* (1540) was an advanced synopsis of Copernicus' book, which he had read in draft.

9. Michael Maestlin (1550–1631), professor of mathematics at Heidelberg, was a strong advocate of Copernicanism, to which he introduced his student Kepler in 1589.

10. David Tost, called "Origanus" (1558–1628), professor of mathematics at Frankfurt, adopted the Tychonic model of the solar system but also accepted the diurnal rotation of the earth.

11. Giordano Bruno of Nola (1548–1600) defended heliocentrism in his *La cena de le ceneri* (1584) and in his *De l'infinito universo e mondi* (1584), although it is now generally agreed that this was not the reason for his execution at the stake on 17 February 1600.

12. Francesco Patrizi of Cherso (1539–1597) seems to have advocated the diurnal but not the annual motion of the earth in his *Nova de universis philosophia* (1593).

13. Galileo (1564–1642) became convinced of Copernicanism sometime before 1597 but did not publicly advocate heliocentrism until 1613 in his *Letters on Sunspots*.

14. Giovanni Antonio Baranzano (1590–1622), who

adopted the name "Redento" when he joined the Barnabite Order, published his *Uranoscopia seu de caelo* in 1617, which explicitly defends Copernicanism.

15. William Gilbert (1540–1603) published his landmark *De magnete* in which the earth is conceived as a large spherical magnet, in 1600.

16. Nicholas Hill (c. 1570–1610) developed nineteen arguments in favor of Copernicanism in Proposition 434 of his *Philosophia Epicurea, Democritana, Theophrastica* (1601).

17. Johannes Kepler (1571–1630), a strong advocate of Copernicanism and the most prominent German astronomer of the day, established modern astronomy with the three laws of planetary motion which still bear his name. The first two of these laws were published in his *Astronomia nova* (1609).

18. Tobias Adami (1581–1643) traveled extensively in Italy in 1611–12, when he began to correspond with and acquire many manuscripts from Campanella. He became Campanella's publisher in Frankfurt after 1617. The additional writings which he promises here appeared the next year (1623) under the title *Philosophia realis epilogistica*, which contained the first published texts of Campanella's *Physiologia, Ethica, Oeconomica,* and *Politica,* the latter with its famous appendix *Civitas Solis* (*The City of the Sun*).

19. Boniface Cardinal Caetani (1567–1617), named a cardinal in 1606 and Bishop of Taranto in 1613, was sympathetic to the new developments in astronomy and opposed the wish of Pope Paul V to condemn Copernicanism as contrary to various passages in Scripture. After Copernicus' book was condemned "until corrected" by the Decree of 5 March 1616, Cardinal Caetani was assigned the task of making the corrections. These were finally published in 1620 in *Monitum Sacrae Congregationis ad Nicholai Copernici lectorem* (English translation in M. Finocchiaro, ed., *The Galileo Affair: A Documentary History* [Berkeley: University of California Press, 1989], 200–202). The corrections turned out to be relatively minor both in extent and in character.

20. In the Introduction to his excellent Italian edition of Campanella's treatise (*Apologia di Galileo* [Turin: UTET, 1968], 21) Luigi Firpo, the leading Italian scholar on this period, has suggested that it is very improbable, if not impossible,

that Cardinal Caetani would have invited Campanella, a convicted prisoner of the Inquisition, to write an evaluation of the Scripture vs. Copernicanism controversy for the Sacred Congregation. As a result he has suggested that Campanella wrote this dedication only after Cardinal Caetani's death on 29 June 1617 and inserted it at this point to enhance the reception of the treatise in Rome. For a detailed discussion of Firpo's interpretation, see the present translator's Introduction to this English edition.

21. It is not clear whether this refers to Cardinal Caetani's role in the deliberations of the Congregation of the Holy Office prior to 5 March 1616 on the orthodoxy of heliocentrism, or to the fact that the Congregation of the Index later assigned to Cardinal Caetani the task of "correcting" Copernicus' *De revolutionibus*, which is mentioned above in note 19.

22. Campanella refers here to the first two of the three articles in his *Questio unica* (1609), which was later given the title *De gentilismo non retinendo*, and published in Paris in 1636 together with his *Atheismus triumphatus* (beginning after page 274) and in 1637 as the Introduction to his *Philosophia realis*.

23. The Fathers of the Church, also called the Apostolic Fathers, were a group of early ecclesiastical writers of the first few centuries after Christ who were held in great esteem, not only because they played a major role in initially establishing Church doctrines, but also because they were considered to be the main conduits through whom the oral tradition of revelation originating from Christ and the Apostles was handed down within the Church. The main Fathers were Gregory Nazianzenus, Basil, John Chrysostom, and Athanasius (from the Greek Church), and Ambrose, Jerome, Augustine, and Gregory the Great (from the Latin Church). Campanella refers to these authors frequently in his *Defense*, following the common theological method of proof by quoted authority used in his day.

24. The Third Book of Esdras, which in the Greek Septuagint Bible was called the First Book of Esdras, is not contained in the canon of the Catholic Bible as defined by the Council of Trent, Session 4 (8 April 1546), but can be found in the Apocrypha of the Old Testament. The passage quoted by Campanella is at 1 Esdras 4:34.

25. This refers to Galileo's *Sidereus nuncius* (1610), in which the mountains on the moon are discussed at length. However the notions that there are also water and an atmosphere on the moon, which are hypothetically mentioned in the same treatise, were soon abandoned by Galileo.

26. Paracelsus (1493–1541), whose real name was Theophrastus Bombastus von Hohenheim, presents his doctrine of multiple human races, corresponding to the four elements, in his *Philosophia ad Athenienses*, II, 11. For an English translation, see *The Hermetic and Alchemical Writings of Aureolus Philippus Theophrastus Bombast*, ed. Arthur Edward White (London: James Elliot, 1894; reprinted by Shambhala Publications, Berkeley, 1976), II, 263 ff.

27. Martin Anton Delrio, S.J. (1551–1608), linguist, philosopher, and scriptural exegete, best known for his *Disquisitionum magicarum libri sex* (Louvain, 1599–1600). The reference is to II, 17, 2, of this work.

28. *Disticha Catonis*, II, 2.

29. This refers to Galileo's announcement of his discovery of four of the satellites of Jupiter in his *Sidereus nuncius* (1610). He called them the "Medicean stars," a name which Campanella also uses for them later in this work.

30. Alessandro Farnese (1468–1549), who was elected pope (Paul III) on 12 October 1534. He initiated the Council of Trent in 1545, and through his legates presided over its Fourth Session (8 April 1546), which established the Decree on the Interpretation of Sacred Scripture, which served later as the main ecclesiastical document on which the 1616 condemnation of Copernicanism was based.

31. See Nicolaus Cardinal Schoenberg's letter to Copernicus (1 November 1536) and Copernicus' dedicatory letter to Pope Paul III, both of which are included as prefaces to the *De revolutionibus orbium coelestium* (1543).

32. In Greek mythology Argus was a creature with one hundred eyes who was always vigilant.

33. Erasmus Reinhold (1511–1553), a student of Copernicus, is best known for his *Prutenicae tabulae coelestium motuum* (1551), which replaced the older Alfonsine Tables.

34. Jan Stade (or Stadius) (1527–1579), an astronomer at Paris, composed the widely used *Ephemerides novae* (1556).

35. Christopher Rothmann (c. 1550–1605), chief astronomer to the Landgrave Wilhelm IV of Hesse–Cassel, was a convinced Copernican and a regular correspondent with Tycho Brahe, against whom he advocated heliocentrism.

36. Campanella's prodigious memory has failed him here. He most likely intended to refer rather to Domenico Maria Novara of Ferrara (1454–1504), who taught astronomy at the University of Bologna from 1483 to 1504, where Copernicus was his student for three and a half years beginning in January of 1497. In his *Apologia di Galileo* (p. 42), Luigi Firpo suggests that Campanella has confused him with Francesco Silvestri (1474–1528), a Dominican theologian at Bologna (1521–1523) and General of the Order of Preachers after 1525, who was not an astronomer.

37. This refers to Giordano Bruno of Nola, who was executed in Rome for heresy in 1600.

38. Johannes Kepler, *Dissertatio cum Nuncio Sidereo . . .* (1610).

39. William Gilbert, *De magnete* (1600).

40. Campanella may have had in mind here, among others, John Field (c. 1525–1587), John Dee (1527–1608), Thomas Digges (?–1595), and Robert Recorde (c. 1510–58).

41. Giovanni Antonio Magini (1555–1617), *Ephemerides coelestium motuum* (1582).

42. Christopher Clavius, S.J. (Christoph Clau) (1537–1612) was the most prominent Jesuit astronomer of the day, whose *In sphaeram Ioannis de Sacro Bosco commentarius* (1570) in its many editions was the standard textbook in astronomy for half a century. Clavius played a major role in the introduction of the Gregorian calendar in 1582. Campanella refers here to the last revision of Clavius' *In sphaeram* in his *Opera mathematica* (1611–12), III, 75.

43. "Apelles" was the pseudonym of Christopher Scheiner, S.J. (1573–1650), whose treatise on sunspots, *De maculis solaribus*, appeared in 1612, and who became the leading authority on sunspots in his day. After 1612 Scheiner and Galileo became increasingly bitter enemies, primarily because of their dispute over priority on the discovery of sunspots.

44. St. Ambrose (c. 340–397), Bishop of Milan, was one of the most important Fathers of the Church. Campanella

quotes him frequently in the *Defense*. The reference is to his *Hexameron*, I, 1, 2–3.

45. Giovanni Pico della Mirandola (1463–1494) was a prominent Renaissance Platonist in the Florentine Academy sponsored by Cosimo de' Medici.

46. Numa Pompilius, King of Rome (715–672 B.C.), was the successor to Romulus and thus the second of the seven legendary kings of Rome.

47. Ovid, *Ex ponto*, III, 3, 44.

48. Pliny the Elder, *Historia naturalis*, XXXIV, 12, 1.

49. Tommaso Casella (1511–72) was a Dominican bishop, theologian, and diplomat.

50. Again, instead of Francesco Maria of Ferrara, Campanella most likely meant Domenico Maria Novara. See note 36.

51. St. Justin Martyr (c. 100–165) was a famous Christian apologist and philosopher who argued against the Greek philosophers as inferior to Christian teachers, a favorite theme for Campanella. The treatise mentioned by Campanella was spuriously attributed to Justin. It is printed among his writings in Migne, *Patrologiae Graeca*, vol. 6; the specific reference is to Questio CXXX, in *PG* 6, 1382.

52. Unlike the previous ten arguments, this eleventh argument in favor of Galileo is not explicitly evaluated by Campanella in chapter 5.

53. Bernard of Clairvaux, O.S.B. (1090–1153), *Apologia ad Guillelmum abbatem* (1125), I, 1–3, and V, 10. In the passage at Romans 10:2, as stated in the Latin Vulgate edition, St. Paul says when speaking of the Jews, "Testimonium enim perhibeo illis quod aemulationem Dei habent, sed non secundum scientiam." ("I give testimony that they have zeal for God but not understanding.")

54. This refers to St. Leo I, also known as Leo the Great, pope from 440–461, and specifically to his *Sermo 59* (Migne, *PL* 54, 337–39).

55. Titus Livius (59 B.C.–17 A.D.), Roman historian.

56. This refers to Campanella's *Atheismus triumphatus* (1636), I, 2–3.

57. See Acts 22:3.

58. Lactantius Firmianus (d. 325) was an early Christian

theologian who opposed natural philosophy as contrary to religion. His denial of the antipodes is located in his *Divinae institutiones*, III, 24.

59. Augustine (354–430), *De civitate Dei*, XVI, 9.

60. The term "antipodes" means literally "what is across from our feet," and was used in medieval and early modern times to refer to both the people and the places located in what we now call the Western hemisphere.

61. Thomas Aquinas (1225–1274), *Summa theologica*, I, 102, 2, ad 4, where he quotes Aristotle (*Meteorologica*, II, 5 [362 b 27]) to the effect that the region under the equator is uninhabitable because of its heat.

62. Albert the Great (Albertus Magnus) (1206–1280) was a philosopher and theologian who taught mostly at Cologne and briefly at Paris (1240–1248) where Thomas Aquinas was his student. See his *De natura locorum*, I, 6–10, for his arguments on this point.

63. Avicenna (Ibn Sina) (980–1037), was a prominent Arabic philosopher and commentator on Aristotle.

64. St. Ephrem (Ephraem, Ephraim) (306/7–372) was a Syriac commentator on the Scriptures.

65. Anastasius Sinaita (Anastasius of Sinai) was a seventh-century Greek saint, theologian, and abbot of the monastery of Mt. Sinai.

66. Moses, Bishop of Syria (Moses bar Cephas) (c. 813–903) was a Syriac theologian and commentator on the Bible and on Aristotle.

67. Thomas Aquinas, *Contra impugnantes Dei cultum et religionem* (Opusculum 19), chap. 11.

68. Terence, *Adelphi*, 98–99.

69. The "Church Militant" refers to the Church in this world, as distinct from the saints in heaven, who constitute the "Church Triumphant."

70. Campanella, *Theologia*, Praefatium, I, 1–4.

71. Thomas Aquinas, *Summa theologica*, II–II, 8, 1, ad 2, and 9, 1 and 4.

72. Fifth Lateran Council, Session 8 (19 December 1513).

73. Thomas Aquinas, *Contra impugnantes Dei cultum et religionem* (Opusculum 19), chap. 12.

74. St. Jerome (c. 340–420) was a Father of the Church,

editor of the Latin Vulgate edition of the Bible, and a commentator on Scripture. The reference is to his *Epistola 70: Ad Magnum*, I, 4.

75. St. Gregory the Great was a Father of the Church and Pope from 590 to 604. The reference is to his *Moralia*, IX, 11, 12 (Migne, *PL* 75, 865–66).

76. John Chrysostom (c. 347–407) was a Father of the Church whose *Homilies* are quoted frequently by Campanella in this work.

77. This is the first line of Aristotle's *Metaphysics* (980 a 23).

78. St. Brigid of Ireland (c. 460–c. 528) is one of the patron saints of Ireland.

79. Leo the Great, *Sermo VII. De ieiunio decimi mensis VII* (Migne, *PL* 54, 185).

80. St. Cyril of Alexandria (early fifth century), *Pro sancta christianorum religione, adversus libros athei Iuliani* (*Contra Iulianum*) (Migne, *PG* 76, 510 ff.).

81. St. Nicephorus (c. 758–829) was Patriarch of Constantinople 806–815.

82. St. Anthony (c. 251–c. 356) was the founder of a Christian monastery at Pisper in Egypt about 305, which became the start of the Catholic monastic system.

83. Chrysostom, *Expositio in Psalm 147*, 3 (Migne, *PG* 55, 482).

84. Ovid, *Metamorphoses*, I, 84–86.

85. Neither of these dialogues are attributed to Plato by modern scholars, although there is some disagreement over the *Epinomis.*

86. Campanella, *Atheismus triumphatus*, VII.

87. Ovid, *Fasti*, I, 297–98; 305–6.

88. Flavius Josephus (37–c. 101), Jewish historian.

89. Philo Judaeus (born c. 25 B.C.), Jewish philosopher and theologian.

90. Berosus (Bèl–usur) (c. 340–c. 270 B.C.), Babylonian historian.

91. An unpublished work by Campanella.

92. Gregory, *Homilia 40*, I, 1 (Migne, *PL* 76, 1078).

93. See Numbers 24:17.

94. Aristotle, *De caelo*, II, 8, 12, 14.

95. Aristotle, *Metaphysics*, XII, 8.

96. Simplicius was an early sixth-century Greek commentator on Aristotle.

97. St. Basil (c. 330–379), Bishop of Caesarea and a Father of the Church, is often quoted by Campanella in this work. This reference is to his *In Hexameron*, I, 11.

98. Ambrose, *Hexameron*, I, 6, 20 and 23.

99. Campanella, *Questiones physiologicae*, 10, 1.

100. Aristotle, *Metaphysics*, XII, 8.

101. Famous mathematicians in Plato's Academy.

102. Thomas Aquinas, *Commentary on Aristotle's Metaphysics*, XII, Lecture 10.

103. According to Copernicus (*De revolutionibus orbium coelestium*, I, 10; II, 1–5) the center of the universe is near, but not exactly coincident with, the sun.

104. Thebit or Tebith (Thabit ibn Qurrah) (836–901) was an Arabic astronomer and translator of Greek mathematical and scientific works.

105. King Alfonso was Alfonso X of Castille (1221–1284), who invited astronomers to prepare what came to be called the Alfonsine Tables.

106. Eugene III, pope from 1145 to 1153. The reference is to Bernard's *De consideratione ad Eugenium III, papam.*

107. Aristotle, *De anima*, I, 1.

108. Richard of St. Victor (d. 1173) was a Scottish theologian and mystic who regarded secular learning as worthless. The reference is to his *Benjamin minor*, or *De praeparatione animi ad contemplationem* (Migne, *PL* 196, 9–10).

109. Plato, *Hippias major*, 282a.

110. See Jeremiah 3:12; 31:54; 33:11.

111. This is a paraphrase of Ecclesiastes 3:11 and Romans 1:20.

112. Clement I (Clemens) (fourth pope, 88–97 A.D.), who knew and worked with St. Paul, and to whom many writings were spuriously attributed. For the reference, see Pseudo–Clement, *Recognitiones*, II, 49–50 (Migne, *PG* 1, 1271–73).

113. John Chrysostom, *Expositio in Psalm 147*, 3–4 (Migne, *PG* 55, 481–83).

114. Pseudo-Clement, *Recognitiones*, I, 8 (Migne, *PG* 1, 1211).

115. Barnabas (first century A.D.), originally named Joseph, was a coworker with St. Paul and is mentioned frequently in Acts.

116. Philo Judaeus, *De vita Mosis*, I, 6.

117. Flavius Josephus, *Antiquitates judaicae*, II, 9, 6; II, 10, 2; IV, 8, 42.

118. On the tabernacle, see Exodus 26:1–37; on the candelabra, see Exodus 25:31–40; on the vestment of Aaron, see Exodus 28:4–43.

119. Augustine, *De Genesi ad litteram imperfectus liber*, 3, 7–9; and *De civitate Dei*, XI, 9, 33. See also Thomas Aquinas, *Summa theologica*, I, 61, 1, ad 1.

120. John Chrysostom, *In Genesim homilia* II, 3 (Migne, *PG* 53, 29).

121. Augustine, *De civitate Dei*, XI, 7; *De Genesi ad litteram*, IV, 26; V, 23, 46; *De Genesi ad litteram imperfectus liber*, 11, 35; 12, 36.

122. Thomas Aquinas, *Summa theologica*, I, 68, 3.

123. Anastasius Sinaita, *Anagogicae contemplationes*. This entire work is an allegorical reading of Genesis.

124. John Chrysostom, *In Genesim homilia* VI, 3 (Migne, *PG* 53, 57–59).

125. Thomas Aquinas, *Summa theologica*, I, 70, 1, ad 3.

126. Fifth Lateran Council, Session 8 (19 December 1513).

127. Averroës (Ibn Rushd) (1126–98) was an Arabic philosopher, astronomer, and legalist who wrote extensive commentaries on Aristotle, thereby gaining the name "the Commentator."

128. Alfarabi (died 950) was an Arabic philosopher, translator, and commentator who attempted to reconcile Platonism and Aristotelianism.

129. Ali Albenragel: this is possibly a reference to Abu'l Hassan Ali, a tenth-century Arabic astronomer at Morocco.

130. Albumasar, or Abubacer (Abu Bakr ibn Thofaïl) (c. 1100–85).

131. Campanella, *Atheismus triumphatus*, XI, 151.

132. *Timaeus*, 27c.

133. John Chrysostom, *In Epistolam ad Romanos*, III, 3 (Migne, *PG* 60, 414).

134. *Phaedo*, 118a.

135. Thomas Aquinas, *Summa contra gentiles*, I, 7–8.

136. Proverbs 9:3 is quoted by Thomas Aquinas to this effect in his *Summa theologica*, I, 1, 5, *sed contra*.

137. Fifth Lateran Council, Session 8 (19 December 1513).

138. The Second Council of Nicaea occurred in 787.

139. This refers to the 219 Articles of Paris condemned by Bishop Ètienne Tempier in 1277 as in part a reprimand to Thomas Aquinas. More specifically Campanella seems to refer here to Article 4, "That the wise men of this world are only the philosophers," and to Article 12, "That one should believe only what is understood directly or is based on what is self-evident." The original condemnation was published in *Chartularium Universitatis Parisiensis*, I, 543–58. A full list of the condemned articles is to be found in P. Mandonnet, *Siger de Brabant et l'averroisme latin au XIIe siècle*, 2 vols. (Louvain, 1908, 1911), II, 175–81.

140. Antonio Berardi della Mirandola (1503–1565) was a teacher of logic at the University of Bologna (1533), Bishop of Caserta (1552), and a commentator on, and staunch defender of, Aristotle.

141. See Plato, *Apology*, 21d; 23b.

142. Augustine, *De doctrina christiana*, II, 40.

143. Thomas Aquinas, *Summa theologica*, I–II, 109, 2 and 3; 110, 1. See also I, 1, 8, ad 2.

144. Jeremiah 2:20.

145. Jerome, *Epistola 57*, 12–13. Pammachius (died c. 409) was a Roman senator who was a friend of Jerome since boyhood.

146. 1 Esdras 10:2, 18–19.

147. Campanella, *De gentilismo non retinendo* (1609).

148. See Genesis 21:10–15.

149. Nehemiah 13:24–27.

150. Thomas Aquinas, *Summa theologica*, I, 1, 8.

151. Campanella's reference here to Pietro Bembo (1470–1547) seems to be an error due to a memory lapse. Apparently what he had in mind was chapter 29 of *Paradossi, cioè sententie fuori del comun parere* (1543) by Ortensio Lando (c.1512–1553). For the textual evidence supporting this

claim suggested by Luigi Firpo, see footnote 75 of his *Apologia di Galileo*, 74–76.

152. Julian the Apostate (31–63) (Roman Emperor 360–363) outlawed Christian schools of rhetoric and grammar even though they studied only the classical pagan authors.

153. Thomas Aquinas, *Contra impugnantes Dei cultum et religionem* (Opusculum 19), chap. 11.

154. Niccolò Machiavelli (1469–1527), whose advocacy of a political philosophy of treachery, deceit, and tyranny in *The Prince* (1513, published 1532) was particularly offensive to Campanella, who often condemns what he calls Machiavellianism.

155. Campanella, *Atheismus triumphatus*.

156. Augustine, *De doctrina christiana*, II, 6.

157. Gregory, Moralia, XV, 13 (Migne, *PL* 75, 1088–89).

158. Augustine, *De Trinitate*, I, 1, 2.

159. Cardinal Cajetan (Tommaso de Vio of Gaeta) (1469–1534), a Dominican philosopher and theologian who was well known for his extensive commentaries (1507–22) on Aquinas' *Summa theologica*.

160. Thomas Aquinas, *Summa theologica*, I, 1, 10 (in the previous sentence); and I, 32, 4 (in this sentence).

161. Augustine, *De Genesi ad litteram*, I, 20.

162. Thomas Aquinas, *Responsio ad magistrum Ioannem de Vercellis de articulis XLII* (Opusculum 10), 18.

163. Thomas Aquinas, *Praeclarae questiones super librum De Trinitate* (Opusculum 70).

164. Thomas Aquinas, *Responsio ad magistrum Ioannem de Vercellis de articulis XLII* (Opusculum 10), Proemium.

165. See chapter 4.

166. Ulisse Albergotti di Arezzo, *Diologo nel quale si tiene, contro l'opinione comune degli astrologi, mathematici e filosofi, la luna esser da sé luminosa e non ricevere il lume dal sole* (1613).

167. The syllogism in brief would be:
The whole human race has descended from Adam (from the Bible). Any humans in the antipodes could not be descendants of Adam. Hence there are no humans in the antipodes (according to the Bible).

168. Lactantius Firmianus, *Divinae institutiones*, III, 24.

169. Augustine, *De civitate Dei*, XVI, 9.

170. Procopius of Gaza (sixth century) was a theologian and scriptural commentator who denied the antipodes because he interpreted the Bible to say that the earth is flat.

171. Dante Alighieri (1265–1321), Italian poet and author of the *Divine Comedy*.

172. St. Isidore of Seville (c.560–636) was a voluminous writer on a wide variety of topics who was widely read in the mediaeval period. See Pseudo–Isidore, *Liber de numeris*, 8 (Migne, *PL* 83, 1298).

173. Xenophanes (570–478? B.C.) was a Greek philosopher who attacked the notion of a plurality of Gods conceived anthropomorphically. The fragments left from Xenophanes do not indicate that he held the view that the earth floats on water.

174. This is an expression of Campanella's own version of geocentrism, which he adopted from Bernardino Telesio (1508–1588), in which the center of the universe is occupied by a motionless, cold, and dark earth around which revolves a fast moving, hot, and luminous sun.

175. Ambrose, *Hexameron*, I, 3.

176. St. Philastrius (died before 397), Bishop of Brescia. Campanella refers here to his *Diversarum hereseon liber* (384), 112.

177. Venerable Bede (672/73–735), Doctor of the Church, whose *Historia ecclesiastica gentis anglorum* (731) is a landmark for the early history of the Church in England. The reference is to his *In Lucam*, IV.

178. Thomas Aquinas, *Summa theologica*, I, 102, 2, ad 4, where he quotes Aristotle [*Meteorologica*, II, 5 (362 b 27)] to the effect that the torrid zone is uninhabitable because of its heat.

179. See Genesis 3:24.

180. Robert Cardinal Bellarmine (1542–1621) was a Jesuit theologian, cardinal (1599), theological adviser to several popes, and a member of the Congregation of the Holy Office. He played a central role in the condemnation of heliocentrism and in the disputed injunction served on Galileo in 1616.

181. Thomas Aquinas, *Responsio ad magistrum Ioannem de Vercellis de articulis XLII* (Opusculum 10), Article 39.

182. Aristotle, *Meteorologica*, II, 1 (354 a 29–30).

183. See chapter 4.

184. Horace, *Epistolae*, II, 1, 83–85.

185. Jerome, *Epistola LXX, Ad Magnum oratorem urbis Romae*, I, 6.

186. The meaning here is that by Campanella's time Thomas Aquinas would have had plenty of reason to wish that genuine theologians would replace those who only slavishly repeat Thomistic teachings.

187. Campanella, *De gentilismo non retinendo* (1609).

188. Thomas Aquinas, *Summa theologica*, I, 1; *Summa contra gentiles*, I, 7–8; *Responsio ad magistrum Ioannem de Vercellis de articulis XLII* (Opusculum 10).

189. Eusebius Pamphili, or Eusebius of Caesarea (c. 260–c. 340), Bishop of Caesarea and the first major writer on Church history. His *Praeparatio evangelica* argued, among other things, that Greek philosophers derived some of their ideas from Jewish sources, especially Moses, a theme which Campanella develops in detail later in the last section of his *Defense* to account for the origin of heliocentrism by Pythagoras.

190. Origen (185–253/4), a Christian apologist, theologian, and biblical exegete.

191. Pseudo-Justin, *Confutatio quorundam dogmatum Aristotelicorum* (Migne, *PG* 6, 1564).

192. This refers to Campanella's *De gentilismo non retinendo* (1609).

193. Aristotle, *Physics*, VIII, 8; and *Metaphysics*, XII, 6.

194. Thomas Aquinas, *Commentary on Aristotle's Metaphysics*, Book XII, Lecture 10.

195. Alexander of Hales (c. 1185–1245), Franciscan philosopher and theologian.

196. St. Vincent Ferrer of Valencia (1346–1419), well known Dominican preacher and missionary.

197. Serafino da Fermo (Seraphinus Acetus de Portis Firmanus) (1496–1540) was the author of *Brevis in Apocalypsim D. Ioannis Apostoli enarratio*, published in *Operum spiritualium* (1570). The reference is to chapter XVI, 556.

198. Origen, *Contra Celsum*, I, 20 (Migne, *PG* 11, 696–97).

199. Campanella, *De gentilismo non retinendo* (1609).

200. Ibid.

201. In Aristotelian cosmology the first four elements (earth, water, air, and fire) compose all objects in the terrestrial world, i.e., everything inside the geocentric orbit of the moon. Everything else in the universe from the sphere of the moon outwards was composed of the immutable "fifth essence" or "quintessence," which undergoes only local motions which are circular and eternal.

202. Theodoretus of Cyrus (c. 393–458), *Questiones in Genesim*, I. See also his commentaries on Psalm 18 and on the *Epistle to the Hebrews*, 19.

203. "A woman clothed with the sun," a phrase taken from Revelations 12:1.

204. Peter Lombard, also called the Master of the Sentences or sometimes simply the Master (c. 1100–c. 1162), was the author of the *Libri IV Sententiarum* (c. 1150), a compilation of texts taken from the Fathers of the Church, especially Augustine, which served for centuries as a basic source of texts and commentaries in the medieval schools. See *Sentences*, II, 14, 4.

205. Ambrose, *Hexameron*, IV, 3, 10; 8, 31.

206. John Philoponus, also call the Grammarian, was a sixth-century Alexandrian philosopher who authored extensive commentaries on Aristotle.

207. Thomas Aquinas, *Summa theologica*, I, 65–74.

208. "Now the fiery sun sets," from Ambrose's Hymn 11, first verse (Migne, *PL* 16, 1476).

209. Aristotle, *De caelo*, II, 7 (289 a 20).

210. Diodorus of Tarsus (died c. 392) was a teacher of John Chrysostom and a biblical commentator who strongly favored the literal over the allegorical reading of Scripture. Only titles and fragments of his writings survive.

211. Thomas Aquinas, *Responsio ad lectorem Venetum de articulis XXXVI* (Opusculum 11), Article 25.

212. Matthew 8:12; 22:13; 25:30.

213. Origen, *In Matthaeum commentariorum series*, 69 (Migne, *PL* 13, 1710–11).

214. John Chrysostom, *In Epistolam ad Thessalonios homilia VII*, 4 (Migne, *PG* 62, 438).

215. Theophylactus of Achrida (c. 1038–1108), Bishop of Bulgaria.

216. Aristotle, *Metaphysics*, I, 8; *De caelo*, II, 9.

217. Ovid, *Metamorphoses*, XV, 148; 239–44; 342–55.

218. Origen, In *Ezechielem homilia IV*, 1 (Migne, *PG* 13, 689).

219. Thomas Aquinas, *Responsio ad lectorem Venetum de articulis XXXVI* (Opusculum 11), Article 24.

220. This probably refers to Gregory of Nyssa (late fourth century), brother of Basil, who was famous for his literal reading of Genesis in his *In Hexameron*.

221. In Latin the noun *caelum* means "the heavens," while the verb *caelere* means "to engrave." For Copernicus' statement of this play on the Latin words (which of course is completely lost in English), see his *De revolutionibus orbium coelestium*, Book I, first paragraph.

222. Valafrido, also called Strabone (c. 808–849), a Swabian theologian, poet, and abbot of Reichenau.

223. Sixtus Senensis (Sisto da Siena) (1520–1569) was a brilliant Jewish convert to Catholicism who was educated by Catharinus and the Dominicans. All of his writings were destroyed except the *Bibliotheca sancta ex praecipuis catholicae Ecclesiae authoribus collecta* (1566), which was published in many editions.

224. Thomas Aquinas, *Responsio ad magistrum Ioannem de Vercellis de articulis XLII* (Opusculum 10), Article 16.

225. Bede, *In Genesim*, 245–53.

226. Peter Lombard, *Libri IV Sententiarum*, II, 14, 5.

227. Thomas Aquinas, *Contra errores Graecorum* (Opusculum 1), Proemium.

228. This refers to the second line of the funereal hymn "Dies irae." The prophet referred to is Joel. See Joel 4:16.

229. Basil, *In Hexameron homilia III*, 5.

230. Thomas Aquinas, *Responsio ad lectorem Venetum de articulis XXXVI* (Opusculum 11), Article 24.

231. "Tophet," a symbol for hell, was the place west of Jerusalem where the city's garbage was burned.

232. Augustine, *De Genesi ad litteram*, XV, 30.

233. Augustine, *De Genesi ad litteram*, I, 10.

234. Lactantius Firmianus, *Divinae institutiones*, III, 24.

235. Aristotle, *Meteorologica*, II, 1 (354 a 29–30).

236. Thomas Aquinas, *Responsio ad magistrum Ioannem*

de Vercellis de articulis XLII (Opusculum 10), Article 18.

237. Thomas Aquinas, *Responsio ad lectorem Venetum de articulis XXXVI* (Opusculum 11), Article 6.

238. Niccolò Antonio Stigliola (Stelliola) of Nola (1546–1624) was one of the earliest and strongest defenders of Copernicanism, for which he was denounced in 1595 to the Inquisition and imprisoned for two years.

239. Heraclitus (fl. c. 500 B.C.), Greek philosopher.

240. Aristarchus of Samos (early third century B.C.), who is generally credited with being the first in the West to formulate a systematic heliocentric astronomy.

241. Philolaus (late fifth century B.C.) was a Pythagorean philosopher at Croton in Southern Italy.

242. Philastrius, *Diversarum hereson liber*, 112.

243. Ulisse Albergotti de Arezzo.

244. Thomas Aquinas, *Summa theologica*, I, 70, 1, ad 3 and ad 5.

245. John Chrysostom, *In Genesim homilia VI*, 3 (Migne, *PG* 53, 57–59).

246. Given his imperfect telescopic instruments, Galileo interpreted what we now know to be the rings of Saturn as two diametrically opposed moons. See Galileo, *Opere*, X, 410.

247. This is the first verse of Ambrose's Hymn 11, "Iam sol recedit igneus."

248. Virgil, *Aeneid*, III, 72.

249. Ambrose, *Hexameron*, I, 6.

250. St. Athanasius (296–373), Bishop of Alexandria and a Doctor of the Church, was active in the theological dispute over Arianism.

251. This refers to Paul's conversion of non-believers in the synagogue in Iconium, as reported in Acts 14:2–3.

252. See Campanella's *Physiologia*, VII, 9; and his *Questiones physiologicae*, 25.

253. Peter Lombard, *Libri IV Sententiarum*, II, 14, 5.

254. Augustine, *De Genesi ad litteram*, II, 5.

255. Thomas Aquinas, *Summa theologica*, I, 70, 1.

256. Pliny the Elder, *Historia naturalis*, II, 4, 4.

257. In fact Galileo attributed an atmosphere only to the moon. See Galileo, *Opere*, III–I, 70.

258. Augustine, *De Genesi ad litteram*, II, 1.

259. Bernardino Telesio (1508–88), *De rerum natura iuxta propria principia libri IX* (1586), II, 19.

260. Galileo, *Letters on Sunspots* (1613), First Letter.

261. Origen, *In Mattheum*, 17, 30; and *In Ieremiam homilia X*, 6.

262. Basil, *In Hexameron homilia* I, 5.

263. Thomas Aquinas, *Summa theologica*, I, 68, 1.

264. Empedocles of Agrigentum (c. 484–424 B.C.) was a Greek natural philosopher who saw the history of the world as composed of the mingling and the separation of the four elements.

265. Plato, *Timaeus*, 28b, 32b–c, 43a, 53b. But see also *Republic*, 616b–e.

266. Marsilio Ficino (1433–1499), *In Timaeum commentarium*, XXIV. Ficino was a strong advocate of Platonism and the central figure at the Platonic Academy in Florence.

267. John Chrysostom, *In Genesim homilia II*, 2 (Migne, *PG* 53, 30).

268. Basil, *In Hexameron homilia III*, 5 (Migne, *PG* 29, 64).

269. Peter Lombard, *Libri IV Sententiarum*, II, 14, 4.

270. Aristotle, *De caelo*, I, 2 (269 b 12–16).

271. Augustine, *De Genesi ad litteram*, II, 11.

272. Thomas Aquinas, *Summa theologica*, I, 68, 1, ad 1.

273. Augustine, *De doctrina christiana*, I, 2; II, 1, 12, 16. See also his *De Genesi ad litteram*, I, 18.

274. Augustine, *De Genesi ad litteram*, II, 4.

275. Basil, *In Hexameron homilia I*, 8.

276. Campanella, *Questiones physiologicae*, XI, 1.

277. Augustine, *De vera religione*, XVIII, 36; *De Genesi ad litteram*, I, 1; IV, 1, 2, 26; V, 3; and *De civitate Dei*, XI, 6–7.

278. Peter Lombard, *Libri IV Sententiarum*, II, 12–15.

279. Augustine, *De civitate Dei*, X, 27.

280. Porphyry (232–c. 300), a neo-Platonic Greek philosopher and associate of Plotinus, was well known for his treatise on Aristotle's *Categories* and for his opposition to Christianity.

281. Campanella, *Questiones physiologicae*, X, 1.

282. Plotinus (205–270) was born in Egypt, settled later

in Rome, and was the most important of the neo-Platonists. His philosophy verged on the mystical. The reference is to *Ennead*, II, 2, 4–5.

283. Xenocrates of Chalcedon (396–315 B.C.) was the second scholarch, or director, of the Academy after Plato.

284. Campanella, *Questiones physiologicae*, XVI, 10–12; XVII; and *Metaphysica*, II, 5, 8.

285. Thomas Aquinas, *Summa theologica*, I, 68, 4.

286. Campanella, *Questiones physiologicae*, XVIII, 1; and *Metaphysica*, II, 5, 4–6.

287. Origen, *In Genesim homilia I*, 2 (Migne, *PG* 12, 148).

288. Basil, *In Hexameron homilia III*, 9.

289. At this point the literal rendition of the Latin text is "between a line and whiteness," which means "between things having nothing in common," when taken in the generalized sense intended by Campanella.

290. The Latin *aether* is related to the Greek *ho aithos*, which means fire. See Augustine, *De Genesi ad litteram imperfectus liber*, 8, and *Confessiones*, XIII, 32.

291. Tycho Brahe (1546–1601), famous Danish astronomer. His *De stella nova* (1573), which announced the new discovery, was reprinted posthumously in his *Astronomiae instauratae progymnasmata* (1602), which Campanella read in 1611. Tycho's "new star" (what we now call a supernova) first appeared in November, 1572, was initially so bright as to be visible during the day, and gradually became invisible to the naked eye about eighteen months later. Its importance at the time was that it provided evidence against the Aristotelian doctrine of the immutability of the heavens.

292. This probably refers to St. Hilary of Poitiers (d. 368).

293. Lancelot Politi (1483–1553) of Siena was a prominent theologian of the Counter-Reformation who adopted the name Ambrogio Caterino (Catharinus) when he joined the Dominican Order in 1517. The reference here is to his *Enarrationes in quinque priora capita libri Geneseos* (1552).

294. Clement of Alexandria (Titus Flavius Clemens) (c. 150–211), an early Greek theologian.

295. This is a reference to Galileo's observations of the

phases of Venus in 1610, as reported in his *Sidereus nuncius.*

296. This refers to the Shiite sect of Islam which became the state religion of Persia in 1502.

297. Thomas Aquinas, *Summa theologica*, I, 47, 3.

298. Democritus of Abdera (c. 460–371 B.C.), a Greek natural philosopher who was the father of atomism.

299. Epicurus of Samos (341–270 B.C.), a Greek atomist and father of hedonism as an ethical theory.

300. Aristotle, *De caelo*, II, 13 (295 b 20–25).

301. The reference is to the Articles of Paris, #27: "Quod Prima Causa non posset plures mundos facere."

302. Aristotle, *De caelo*, I, 8–9.

303. Galileo, *Letters on Sunspots*, Third Letter. *Opere*, V, 220.

304. Campanella, *Questiones physiologicae*, X, 3; XXX-IV, 3.

305. Johannes Kepler, *Dissertatio cum Nuncio Sidereo*, in *Opera*, II, 500.

306. Aristotle, *Metaphysics*, XII, 8 (1074 a 1–15).

307. The parable of the talents is found in Matthew 25:14–30, and in Luke 19:12–27.

308. Gregory the Great, *Homilae in Ezechielem*, I, 7, 5 (Migne, *PL* 76, 842).

309. Campanella, *Questiones physiologicae*, X, 1–2; and *Metaphysica*, XI, 8, 2 and 6; XI, 15, 1–4

310. Campanella, *Physiologia*, III.

311. St. Gelasius I (pope, 492–496), whose *Catalogue* attempted to list the authentic writings of the Fathers of the Church.

312. Gratianus, the Canonist, *Decretum, emendatum et notationibus illustratum una cum glossis, Gregorii XIII Pont. Max. iussu editum*. Rome: In Aedibus Populi Romani, 1584) Part I, Dist. XV, Chap. III: *Quae concilia sancta Romana Ecclesia suscipiat*, 51.

313. Campanella did not write a specific evaluation of the Eleventh Argument.

314. Thomas Aquinas, *Summa theologica*, I, 68, 4.

315. Eusebius, *Historia ecclesiastica*, VI, 19, 2–4.

316. See Clement of Alexandria, *Stromata*, I, 14 (Migne, *PG* 8, 760) and Ambrose, *Epistolae*, VI, 1. For the commenta-

tor on Ambrose, see Migne, *PL* 16, 1095, notes 7 and 8.

317. Gabriele Barrio (c. 1510–c. 1577) was a learned historian and geographer. The reference is to his *De antiquitate et situ Calabriae* (Rome: De Angelis, 1571), IV, 9–11.

318. Campanella, *Metaphysica*, XVI, 7.

319. Diogenes Laertius (third century A.D.) was the author of the not fully reliable *Lives, Opinions, and Sayings of the Philosophers*.

320. Plutarch (c. 45–125 A.D.) was a Greek biographer, essayist, and author of *Parallel Lives of Illustrious Greeks and Romans*.

321. Galen (Claudius Galenus) (c. 138–c. 201), Greek anatomist, physiologist, and intellectual biographer.

322. Again Campanella mentions Francesco Silvestri when he apparently meant Domenico Maria Novara. See note 36.

323. This appears to be a reference to Timaeus of Locri in Plato's *Timaeus* (34a). He does speak there of the earth having a uniform rotation on its axis but does not specify that it is diurnal.

324. This is a reference to Copernicus' "third motion" of the earth, introduced by him to hold the axis of the earth in the same orientation relative to the universe as a whole as it rotates around the sun. This was seen to be an unnecessary fictional motion after Aristotle's solid spheres were eliminated from astronomy by the end of the sixteenth century.

325. Campanella, *Questiones physiologicae*, XI, 1.

326. Thomas Aquinas, *Commentary on Aristotle's Metaphysics*, XII, Lecture 10.

327. Pherecydes of Siros, in the Cyclades (sixth century B.C.) was a Greek philosopher who was apparently either a teacher or an associate of Pythagoras.

328. Dionysius the Carthusian (Leuwis de Rickel) (1402–1471) was a theologian, mystic, and author of *Contra Alcoranum et sectam Machometicam* (1533), to which Campanella refers here.

329. The text at Job 20:17 mentions rather "rivers of cream and honey."

330. This is a general reference to Bellarmine's *Disputationes de controversiis christianae fidei*, 3 volumes (1586–1593),

and perhaps also a special appeal for the favorable consideration of Cardinal Bellarmine, the most powerful member of the Congregation of the Holy Office then sitting in judgment on Copernicanism, for which the *Defense* was written.

BIBLIOGRAPHY

I: EDITIONS OF THE *DEFENSE*

Campanella, O.P., Tommaso. *Apologia pro Galileo, mathematico florentino, ubi disquiritur, utrum ratio philosophandi, quam Galileus celebrat, faveat sacris scripturis, an adversetur.* Frankfurt: Impensis Godefridi Tampachii, Typis Erasmi Kempfferi, 1622.

———. *The Defense of Galileo.* For the First Time Translated and Edited, with Introduction and Notes, by Grant McColley. *Smith College Studies in History* 22 (1937), nos. 3–4. [not a reliable translation]

———. *Apologia di Galileo.* A cura di Luigi Firpo. Turin: Unione Tipografico-Editrice Torinese, 1968. [Italian translation plus a facsimile reprint of the 1622 first edition]

———. *Apologia per Galileo.* A cura di Salvatore Femiano. Milan: Marzorati Editore, 1971. [Italian translation plus a corrected reprinting of the 1622 first edition]

II: STUDIES OF CAMPANELLA

Amabile, Luigi. *Fra Tommaso Campanella: La sua congiura, i suoi processi e la sua pazzia.* 3 vols. Naples: Morano, 1882.

Amerio, Romano. *Campanella.* Brescia: "La Scuola," 1947.

———. "Galilei e Campanella." In *Nel terzo centenario della morte di G. Galilei, 299–325.* Milan: Vita e Pensiero, 1942.

———. *Il sistema teologico di Tommaso Campanella.* Milan and Naples: Riccardo Ricciardi, 1972.

Blanchet, Léon. *Campanella.* Paris: Alcan, 1920; New York: Burt Franklin, 1964.

Bock, Gisela. *Thomas Campanella.* Tübingen: Niemeyer, 1974.
Bonansea, Bernardino M.. "Campanella's Defense of Galileo."
In *Reinterpreting Galileo,* ed. William A. Wallace.
Washington, D.C.: Catholic University of America Press,
1986. 205–39.
————. *Tommaso Campanella: Renaissance Pioneer of
Modern Thought.* Washington, D.C.: Catholic University
of America Press, 1969.
Calogero, Giuseppe. *Tommaso Campanella, prometeo del ri-
nascimento.* Messina: Samperi, 1961.
Corsano, Antonio. "Campanella e Galileo." *Giornale critico
della filosofia italiana* 44 (1965): 313–32.
————. *Tommaso Campanella.* Revised edition. Bari: Laterza,
1961.
Di Napoli, Giovanni. *Tommaso Campanella, filosofo della
restaurazione cattolica.* Padua: Cedam, 1947.
Femiano, Salvatore. *Lo spiritualismo di Tommaso Campanella.*
2 vols. Naples: Instituto Editoriale del Mezzogiorno,
1965. Revised edition in one volume: *La metafisica di
Tommaso Campanella.* Milan: Marzorati, 1968.
————. *Studi sul pensiero di Tommaso Campanella.* Bari:
Editoriale Universitaria, 1973.
Firpo, Luigi. *Bibliografia degli scritti di Tommaso Campanella.*
Turin: V. Bona, 1940.
————. *Ricerche campanelliane.* Florence: G.C. Sansoni Edi-
tore, 1947.
————. *Il supplizio di Tommaso Campanella: Narrazoni, Doc-
umenti, Verbali delle torture.* Rome: Salerno Editrice,
1985.
Grillo, Francesco. *Tommaso Campanella in America: A Critical
Bibliography and a Profile.* New York: S.F. Vanni, 1954.
A Supplement to the Critical Bibliography. New York: S.
F. Vanni, 1957.
Langford, James R.. "Science, Theology, and Freedom." In *On
Freedom,* ed. by Leroy S. Rouner, Notre Dame, Ind.:
University of Notre Dame Press, 1989.
————. "Campanella on Scientific Freedom." *Reality* 11
(1963): 133–50.
Nicotra, Alfio, and Antonietta. *Tommaso Campanella.* Flo-
rence and Catanzara: Mauro, 1948.

Reeves, George C. *The Philosophy of Tommaso Campanella, with Special Reference to His Doctrine of the Sense of Things and of Magic.* PhD. diss., Indiana University, 1935.

Rossi, Mario M.T. *Campanella, metafisico.* Florence: Carpigiani e Zipoli, 1923.

Testa, Aldo T. *Campanella.* Milan: Garzanti, 1941.

Valeri, Nino. *Campanella.* Rome: Formiggini, 1931.

Treves, Paolo. *La filosofia politica di Tommaso Campanella.* Bari: Laterza, 1930.